A Basic Mathematics Work Book

A Basic Mathematics Workbook

Over 3000 Examples for Practice in the Basic Skills of Arithmetic, Algebra and Geometry.

by Elizabeth Miller, M.A.

HAMILTON PUBLISHING

© Hamilton Publishing Co. Ltd.

First Published 1982 by
The Hamilton Publishing Company Limited
12 Colvilles Place
Kelvin
East Kilbride
Glasgow

All rights reserved. No part of this publication may be reproduced, stored in a retrieval system, or transmitted in any form or by any means electronic, mechanical, photocopying, recording or otherwise, without the prior permission of The Hamilton Publishing Company Limited.

I.S.B.N. 0-946164-07-X

Illustrations by Craig Peacock

Produced in Great Britain by
Thomson Litho East Kilbride Scotland

Foreword

A text-book written mainly for pupils in the early stages of mathematics.

Can be usefully employed for examination revision for first year examinations and as an excellent source of questions for class work and home work.

The workbook can also be profitably used for older pupils who are not studying for Certificate Examinations.

Contains over 3,000 examples for practice.

Contents

Section 1: Arithmetic

Exercise 1	Notation: figures → words.	
Exercise 2	Notation: words → figures.	
Exercise 3	Addition of whole numbers.	
Exercise 4	Arranging whole numbers in ascending order.	
Exercise 5	Arranging whole numbers in descending order.	
Exercise 6	Subtraction of whole numbers.	
Exercise 7	Multiplication of whole numbers.	
Exercise 8	Division of whole numbers.	
Exercise 9	Multiplication by 10 and multiplication by 100.	
Exercise 10	Division by 100.	
Exercise 11	Addition of decimals.	
Exercise 12	Subtraction of decimals.	
Exercise 13	Manipulation of decimals. Interpretation needed.	
Exercise 14	Manipulation of decimals. Interpretation needed.	
Exercise 15	Decimals → vulgar fractions.	
Exercise 16	Decimals → mixed numbers.	
Exercise 17	Multiplication of decimals.	
Exercise 18	L.C.M.	
Exercise 19	Reduction of fractions to fractions with common denominators.	
Exercise 20	Addition of fractions.	
Exercise 21	Subtraction of fractions.	
Exercise 22	Simplification of fractions.	
Exercise 23	Arrangement of fractions in ascending order.	
Exercise 24	Multiplication of fractions.	
Exercise 25	Division of fractions.	
Exercise 26	Fractions of concrete quantities.	
Exercise 27	Fractions→percentages.	
Exercise 28	Percentages → fractions.	
Exercise 29	Mixed numbers → percentages.	
Exercise 30	Percentages to fractions.	
Exercise 31	Decimals → percentages.	
Exercise 32	Percentages to decimals.	
Exercise 33	Percentages of concrete quantities.	
Exercise 34	Manipulation. Interpretation needed.	
Exercise 35	Metric conversions. Length.	
Exercise 36	Metric conversions. Length.	
Exercise 37	Metric additions and conversions. Length.	
Exercise 38	Area of rectangle.	
Exercise 39	$D = S \times T$.	
Exercise 40	$S = \dfrac{D}{T}$	
Exercise 41	$T = \dfrac{D}{S}$	

Exercise 42 Profit and loss.
Exercise 43 Volume of cuboid.
Exercise 44 Capacity.
Exercise 45 Conversion of speeds.
Exercise 46 Miscellaneous metric problems.
Exercise 47 Metric addition and conversion. Weight.
Exercise 48 Metric subtraction and conversion. Weight.
Exercise 49 Miscellaneous metric problems. Interpretation needed.
Exercise 50 Hire purchase.
Exercise 51 Profit and loss.
Exercise 52 Profit and loss.
Exercise 53 Simple interest.
Exercise 54 Wages calculations.

Section 2: Algebra

Exercise 1 Numerical substitution.
Exercise 2 Numerical substitution.
Exercise 3 Numerical substitution.
Exercise 4 Collection of like terms.
Exercise 5 Numerical substitution.
Exercise 6 Multiplication of algebraic terms.
Exercise 7 Division of algebraic terms.
Exercise 8 Symbolic expression.
Exercise 9 Symbolic expression.
Exercise 10 Metric conversion, using symbols.
Exercise 11 Metric conversion, using symbols.
Exercise 12 Metric conversion, using symbols.
Exercise 13 Formula substitution.
Exercise 14 Formula substitution.
Exercise 15 Formula substitution.
Exercise 16 Collection of like terms.
Exercise 17 Collection of like terms.
Exercise 18 Algebraic multiplication and division.
Exercise 19 True and false statements.
Exercise 20 True and false statements.
Exercise 21 Simple equations.
Exercise 22 Simple equations.
Exercise 23 Simple equations.
Exercise 24 Simple equations.
Exercise 25 Simple equations.
Exercise 26 Numerical substitution.
Exercise 27 Numerical substitution.
Exercise 28 Simple problems.
Exercise 29 Algebraic multiplication and division.

Exercise 30 True and false statements.
Exercise 31 Simple equations.
Exercise 32 Symbolic expression.
Exercise 33 Symbolic expression.
Exercise 34 Algebraic simplification.
Exercise 35 Simple equations.

Section 3: Some Practice with Simple Geometry Problems.

Nos. 1–510.

Section 4: Geometry Exercises.

Exercise 1 Line drawing.
Exercise 2 Angle sizes.
Exercise 3 Line segments.
Exercise 4 Angles of a triangle.
Exercise 5 Bearings.
Exercise 6 Perimeter of rectangle.
Exercise 7 Perimeter and area of square.
Exercise 8 Parallel lines.
Exercise 9 Drawing of specified triangles.
Exercise 10 Drawing of specified triangles.
Exercise 11 Angle drawing.
Exercise 12 Four-sided figures.
Exercise 13 The angles of the triangle.
Exercise 14 Complements of angles.
Exercise 15 Supplements of angles.
Exercise 16 Drawing of four-sided figures.
Exercise 17 Miscellaneous problems.
Exercise 18 Cubes and cuboids.
Exercise 19 Miscellaneous problems.
Exercise 20 Miscellaneous problems.

Section 5: 500 Miscellaneous Questions.

SECTION 1

ARITHMETIC

Exercise 1

Write, in words:

1. 231	2. 46	3. 2846	4. 12	5. 16 241
6. 725	7. 19	8. 44 642	9. 191	10. 46 451
11. 59	12. 23 468	13. 846	14. 8294	15. 1 234 689
16. 3926	17. 2	18. 34	19. 346 832	20. 14 247

Exercise 2

Write in number form:

1. One hundred and twelve.
2. Forty nine.
3. One million four hundred thousand.
4. Nine.
5. Three thousand and nineteen.
6. Twenty three.
7. Eighty four thousand, three hundred and one.
8. Twenty four hundred.
9. Nineteen.
10. Three thousand, two hundred and forty six.
11. Three hundred and four.
12. Fifty six.
13. Four thousand and nine.
14. One million and one.
15. Twenty thousand two hundred.
16. Ninety nine.
17. Two hundred and one.
18. Three hundred and four.

19. Half a million.
20. One quarter of a million.

Exercise 3

Add the following numbers:

1. 231, 16 and 482.
2. 34, 2618 and 79.
3. 1, 14 and 1235.
4. 76, 760 and 7600.
5. 23, 213 and 27.
6. 3, 31 and 1234.
7. 4, 81, 237 and 19.
8. 2, 29, 326 and 1.
9. 324, 1, 49 and 23.
10. 34, 276, 2834 and 3.
11. 3, 39, 374 and 29.
12. 47, 471, 4171 and 12.
13. 3, 24, 241 and 6.
14. 36, 312, 3141 and 14.
15. 3469 and 2183.
16. 231 461 and 12 146.
17. 4174 and 1389.
18. 1, 11, 123 and 64.
19. 32, 46, 81 and 104.
20. 746 189 and 203.

Exercise 4

Arrange in **ascending** order:

1. 46, 2, 11, 1, 19.
2. 24, 16, 29, 4, 82.
3. 1, 462, 23, 89, 2.
4. 3, 33, 23, 41, 0.
5. 12, 122, 14, 29, 3.
6. 27, 46, 12, 14, 23.
7. 3, 8, 4, 1, 17.
8. 3, 8, 2, 9, 11, 12.
9. 18, 17, 16, 15.
10. 11, 10, 9, 13, 17, 2.

Exercise 5

Arrange in **descending** order:

1. 7, 4, 17, 19.
2. 231, 8, 14, 16, 0.
3. 123, 312, 7, 8, 91.
4. 7, 74, 17, 18, 2.
5. 7, 19, 93, 18, 6, 2.
6. 2314, 2013, 6, 813.
7. 14, 104, 401, 132, 14.
8. 0, 18, 23, 17, 14.
9. 13, 14, 12, 123, 89.
10. 2316, 34, 181, 21.

Exercise 6

In each case take the smaller number from the larger number:

1. 234, 46.
2. 389, 1924.
3. 37, 34.
4. 128, 896.
5. 3714, 14.
6. 314, 41.
7. 918, 874.
8. 33, 310.
9. 913, 912.
10. 84, 48.
11. 346, 347.
12. 919, 191.

| 13. | 1342, 968. | 14. | 321, 1927. | 15. | 1684, 73. | 16. | 384, 1382. |
| 17. | 1974, 198. | 18. | 394, 93. | 19. | 2641, 1642. | 20. | 234, 2034. |

Exercise 7

Work answers for:

1.	23×261.	2.	27×301.	3.	406×41.
4.	71×710.	5.	18×108.	6.	321×31.
7.	384×31.	8.	1010×31.	9.	301×101.
10.	1710×23.	11.	3812×29.	12.	721×41.
13.	63×36.	14.	131×21.	15.	714×11.
16.	21×12.	17.	234×49.	18.	28×32.
19.	2146×13.	20.	1003×13.		

Exercise 8

In each case state the **quotient** and the remainder, if any.

1.	$342 \div 4$.	2.	$3215 \div 5$.	3.	$264 \div 8$.
4.	$291 \div 5$.	5.	$1648 \div 3$.	6.	$2846 \div 9$.
7.	$3205 \div 5$.	8.	$8321 \div 7$.	9.	$346 \div 8$.
10.	$3214 \div 3$.				

Exercise 9

Multiply the following numbers by 10.

1.	12.	2.	34·6	3.	8·1
4.	72	5.	7·81	6.	19
7.	184	8.	10·1	9.	13·2
10.	11·1	11.	34	12.	79·1
13.	73	14.	17·3	15.	1·79
16.	18	17.	32·14	18.	347
19.	12·121	20.	9		

When you have completed **Exercise 9**, use the numbers given in 1 to 20 and multiply each number by 100.

Exercise 10

Using the numbers from 1 to 20 in **Exercise 9, divide** each number by 100.

Exercise 11

Add together:

1. 23·1 and 46·2.
2. 103 and 0·021.
3. 41·6 and 0·03.
4. 29 and 32·1.
5. 0·08 and 1·24.
6. 3·74 and 0·084.
7. 43·4 and 28·6.
8. 32·9 and 18·4.
9. 46 and 28·1.
10. 483, 28·1 and 7·04.
11. 2·31, 3·42 and 6·89.
12. 31·04 and 126·9.
13. 28·4 and 32·01.
14. 4·61, 0·21 and 1·24.
15. 2·14, 16·4 and 1·82.
16. 32·8, 28·6 and 2.
17. 1·5 and 106·23.
18. 0·02, 0·04 and 1·21.
19. 1·46, 1·234 and 2·8.
20. 2·3, 1·21 and 2·64.

Exercise 12

1. From 11·2 take 8·34.
2. From 17·4 take 1·02.
3. Take 32·1 from 42.
4. Take 89·4 from 91·2.
5. From 7 take 3·24.
6. Take 2·41 from 8·4.
7. Take 7·91 from 8.
8. Take 1·21 from 2·1.
9. From 8·46 take 6·48.
10. From 8·72 take 8·27.
11. Take 3·402 from 4·6.
12. Take 2·13 from 2·84.
13. From 8·27 take 1·829.
14. Take 2 from 17·1.
15. Take 2·04 from 2·1.
16. Take 1·3 from 2·04.
17. From 172·4 take 163·8.
18. Take 2·04 from 2·862.
19. From 6·9 take 2·24.
20. Take 17·3 from 20.

Exercise 13

Multiply the first number by 10, multiply the second number by 100 and add your results, in questions 1-20. Perform the divisions in questions 21-30.

1. 27·4, 17.
2. 1·21, 11.
3. 4·32, 46.
4. 49·2, 97.
5. 6·3, 72·1.
6. 8·9, 7.
7. 4, 5·2.
8. 4·1, 0·8.
9. 2·4, 2·04.
10. 7·12, 7·3.
11. 8, 11.
12. 34, 43.
13. 3·2, 1·8.
14. 1·8, 1·08.
15. 71, 17.
16. 89, 8·9.
17. 1·4, 46.
18. 5·05, 1·25.
19. 4·6, 2·64.
20. 8·7, 8·07.
21. Divide 1·2 by 0·2.
22. Divide 1·4 by 0·02.
23. Divide 8·4 by 0·3.
24. Divide 2·1 by 0·7.
25. Divide 2·26 by 0·002.
26. Divide 1·48 by 0·004.
27. Divide 5·235 by 5.
28. Divide 12·32 by 11.
29. Divide 1·46 by 0·2.
30. Divide 1·2342 by 3.

Exercise 14

1. From 7·43 take 1·29. Multiply your result by 1000.
2. Take 24·6 from 30. Multiply your result by 100.
3. Add 1·32 and 2·46. Multiply your result by 10 000.
4. Multiply 43 by 2·1 and multiply your result by 100.
5. Add 3·4, 5·6, 23 and 48·1. Divide your answer by 1000.
6. Add 34 × 10 to 24·1 × 100.
7. Take 34·2 × 10 from 1000.
8. From 8·7 × 3 take 2·4 × 0·03.
9. From 7·4 take 18 × 0·01.
10. Take 12·1 from 1·18 × 100.
11. Take 1·2 from 2 × 0·01
12. Add 1·23 × 10 to 100 × 0·06.
13. Take 0·7 from 100 × 0·032.
14. Add 1·23 and 2·32. Multiply your result by 1000.
15. Multiply 2·32 by 2·1. Add 1·46 to your answer.
16. Add 1·2, 2·4 and 6·8. Divide your result by 0·02.
17. Take 100 × 1·24 from 10 × 22·3.
18. Multiply 1·42 by 2·3. Take 0·83 from your answer.
19. Add 10 × 0·01 to 100 × 0·04.
20. From 17·4 take 15·08. Multiply the result by 1000.
21. Take 42·6 × 10 from 100 × 53·9.
22. Add 3·4, 4·62 and 5·28. Multiply your answer by 100.
23. From 74·1 ÷ 10 take 0·003 × 100.
24. Add 1·21 and 2·34 and take the result from 7.
25. Take 4·32 × 10 from 98·9.
26. Multiply 4·2 by 1·1 and add 73·4 to your result.
27. From 8 × 2·2 take 14·34.
28. Multiply 1·7 by 10 and add 2·3 × 100 to your result.
29. Divide 18·2 by 100 and multiply your answer by 14.
30. Add 1·21 × 2 to 14·86 × 3.

Exercise 15

Write as vulgar fractions in their **lowest terms:**

1. 0·02
2. 0·18
3. 0·6
4. 0·64
5. 0·92
6. 0·24
7. 0·76
8. 0·38
9. 0·78
10. 0·16
11. 0·1
12. 0·001
13. 0·002
14. 0·4
15. 0·42
16. 0·46
17. 0·48
18. 0·96
19. 0·72
20. 0·65

Exercise 16

Write, as mixed numbers:

1. 1·12
2. 7·4
3. 12·8
4. 6·24
5. 14·2
6. 12·06
7. 10·01
8. 2·46
9. 4·96
10. 6·84
11. 2·2
12. 3·86
13. 2·92
14. 1·52
15. 1·62
16. 3·12
17. 2·82
18. 9·6
19. 3·45
20. 4·32

Exercise 17

Work answers for:

1. 1·2 × 0·4.
2. 3·22 × 0·21.
3. 4·6 × 1·2.
4. 3·24 × 0·08.
5. 7·21 × 1·2.
6. 0·04 × 1·2.
7. 8·6 × 0·3.
8. 44·1 × 1·1.
9. 2·24 × 0·01.
10. 1·68 × 2·3.
11. 11·2 × 2·4.
12. 13·8 × 1·3.
13. 82·1 × 0·8.
14. 72·6 × 0·7.
15. 86·4 × 1·4.
16. 2·7 × 0·5.
17. 3·42 × 0·05.
18. 8·82 × 0·005.
19. 1·6 × 1·6.
20. 2·4 × 0·24.

Exercise 18

What is the **lowest** number that each of the numbers in the following exercises divides into **without a remainder:**

1. 2, 3, 4.
2. 2, 3, 6.
3. 3, 4, 6.
4. 2, 4, 8.
5. 2, 6, 12.
6. 2, 3, 12.
7. 3, 4, 12.
8. 4, 6, 12.
9. 2, 3, 8.
10. 3, 8, 10.
11. 3, 6, 8.
12. 6, 8, 12.
13. 6, 12, 24.
14. 2, 5, 15.
15. 2, 6, 15.

16. 3, 8, 15. **17.** 8, 12, 15. **18.** 2, 12, 20.
19. 1, 4, 8, 16. **20.** 3, 4, 8, 16.

Exercise 19

1. Express with denominator 12:
   $\frac{1}{2}; \frac{1}{3}; \frac{2}{3}; \frac{1}{6}; \frac{5}{6}$

2. Express with denominator 40:
   $\frac{1}{5}; \frac{2}{5}; \frac{1}{10}; \frac{3}{10}; \frac{7}{10}$

3. Express with denominator 18:
   $\frac{2}{3}; \frac{1}{9}; \frac{5}{6}; \frac{1}{6}; \frac{4}{9}$

4. Express with denominator 30:
   $\frac{1}{2}; \frac{1}{3}; \frac{2}{5}; \frac{1}{6}; \frac{3}{10}$

5. Express with denominator 50:
   $\frac{1}{5}; \frac{3}{10}; \frac{3}{25}; \frac{1}{2}; \frac{7}{10}$

6. Express with denominator 24:
   $\frac{1}{3}; \frac{5}{8}; \frac{1}{6}; \frac{3}{4}; \frac{5}{6}$

7. Express with denominator 60;
   $\frac{7}{12}; \frac{1}{5}; \frac{7}{30}; \frac{1}{2}; \frac{5}{6};$

8. Express with denominator 21:
   $\frac{1}{3}; \frac{2}{3}; \frac{1}{7}; \frac{3}{7}; \frac{5}{7}$

9. Express with denominator 28:
   $\frac{1}{2}; \frac{1}{4}; \frac{3}{4}; \frac{3}{7}; \frac{5}{7}$

10. Express with denominator 100:
    $\frac{1}{10}; \frac{3}{4}; \frac{1}{25}; \frac{3}{20}; \frac{7}{50}$

Exercise 20

Add together:

1. $\frac{1}{2}$ and $\frac{1}{4}$.
2. $\frac{1}{2}$ and $\frac{1}{2}$.
3. $\frac{1}{4}$ and $\frac{1}{4}$.
4. $\frac{1}{2}$ and $\frac{3}{4}$.
5. $\frac{3}{4}$ and $\frac{3}{4}$.
6. $\frac{1}{2}$ and $\frac{1}{3}$.
7. $\frac{1}{2}$ and $\frac{1}{8}$.
8. $\frac{1}{2}$ and $\frac{1}{6}$.
9. $\frac{1}{3}$ and $\frac{1}{4}$.
10. $\frac{1}{4}$ and $\frac{1}{8}$.
11. $\frac{1}{4}$ and $\frac{1}{6}$.
12. $\frac{1}{6}$ and $\frac{1}{6}$.
13. $\frac{1}{6}$ and $\frac{5}{6}$.
14. $\frac{1}{4}$ and $\frac{1}{16}$.
15. $\frac{1}{16}$ and $\frac{1}{32}$.
16. $\frac{1}{4}$ and $\frac{1}{10}$.
17. $\frac{1}{10}$ and $\frac{5}{6}$.
18. $\frac{1}{6}$ and $\frac{1}{8}$.
19. $\frac{1}{8}$ and $\frac{1}{12}$.
20. $\frac{1}{12}$ and $\frac{1}{2}$.

Exercise 21

Take the first fraction from the second.

1. $\frac{1}{2}, \frac{3}{4}$
2. $\frac{1}{8}, \frac{7}{8}$
3. $\frac{1}{6}, \frac{5}{6}$
4. $\frac{1}{4}, \frac{3}{4}$
5. $\frac{1}{10}, \frac{7}{10}$
6. $\frac{1}{10}, \frac{1}{5}$
7. $\frac{1}{8}, \frac{1}{4}$
8. $\frac{1}{10}, \frac{1}{4}$
9. $\frac{1}{2}, \frac{7}{8}$
10. $\frac{1}{2}, \frac{11}{12}$
11. $\frac{1}{8}, \frac{1}{6}$
12. $\frac{1}{9}, \frac{1}{3}$
13. $\frac{1}{8}, \frac{1}{2}$
14. $\frac{1}{6}, \frac{1}{2}$
15. $\frac{1}{5}, \frac{1}{2}$

16. $\frac{1}{16}, \frac{1}{8}$ 17. $\frac{1}{20}, \frac{1}{10}$ 18. $\frac{1}{5}, \frac{3}{4}$
19. $\frac{1}{3}, \frac{2}{3}$ 20. $\frac{1}{10}, \frac{19}{20}$

Exercise 22

Work answers for:

1. $\frac{1}{2} + \frac{1}{3} - \frac{1}{6}$
2. $\frac{1}{3} + \frac{2}{3} - \frac{5}{6}$
3. $\frac{1}{4} + \frac{1}{2} - \frac{1}{8}$
4. $\frac{1}{6} - \frac{1}{8} + \frac{1}{3}$
5. $\frac{1}{10} - \frac{3}{10} + \frac{5}{6}$
6. $\frac{1}{5} + \frac{3}{5} - \frac{1}{15}$
7. $\frac{1}{10} + \frac{1}{20} + \frac{1}{30}$
8. $\frac{1}{8} + \frac{7}{8} - \frac{1}{6}$
9. $\frac{1}{2} + \frac{1}{8} + \frac{7}{8}$
10. $\frac{1}{3} + \frac{1}{3} + \frac{1}{3}$
11. $\frac{1}{2} + \frac{1}{3} + \frac{7}{8}$
12. $1 - \frac{1}{2} - \frac{1}{4}$
13. $3 + \frac{1}{2} - \frac{1}{4}$
14. $2\frac{1}{2} + 1\frac{1}{3}$
15. $1\frac{7}{8} - \frac{1}{8}$
16. $3\frac{1}{4} + 2\frac{1}{8}$
17. $3\frac{1}{2} - 1\frac{3}{4}$
18. $1\frac{2}{3} - 1\frac{1}{6}$
19. $2\frac{1}{4} + 3\frac{1}{2} - 4\frac{1}{8}$
20. $3\frac{1}{2} + 2\frac{1}{2} + 5\frac{1}{8}$

Exercise 23

Place the fractions in each of the following questions in **ascending** order.

1. $\frac{1}{2}; \frac{2}{3}; \frac{1}{4}; \frac{5}{6}$
2. $\frac{1}{3}; \frac{1}{6}; \frac{1}{4}; \frac{2}{9}$
3. $\frac{1}{10}; \frac{1}{15}; \frac{1}{6}; \frac{7}{40}$
4. $\frac{2}{3}; \frac{7}{8}; \frac{1}{12}; \frac{5}{6}$
5. $\frac{5}{9}; \frac{11}{18}; \frac{2}{3}; \frac{31}{36}$
6. $\frac{1}{4}; \frac{1}{5}; \frac{1}{6}; \frac{1}{12}$
7. $\frac{2}{3}; \frac{7}{20}; \frac{11}{40}; \frac{9}{20}$
8. $\frac{3}{5}; \frac{9}{20}; \frac{7}{10}; \frac{1}{4}$
9. $\frac{2}{3}; \frac{7}{8}; \frac{7}{10}; \frac{5}{6}$
10. $\frac{3}{5}; \frac{5}{8}; \frac{3}{4}; \frac{7}{16}$
11. $\frac{5}{6}; \frac{3}{4}; \frac{11}{12}; \frac{19}{24}$
12. $\frac{5}{6}; \frac{1}{2}; \frac{2}{3}; \frac{7}{8}$
13. $\frac{7}{10}; \frac{3}{5}; \frac{9}{20}; \frac{11}{15}$
14. $\frac{1}{2}; \frac{2}{3}; \frac{3}{4}; \frac{4}{5}$
15. $\frac{7}{8}; \frac{11}{16}; \frac{13}{24}; \frac{3}{4}$
16. $\frac{4}{5}; \frac{7}{10}; \frac{19}{20}; \frac{21}{40}$
17. $\frac{4}{5}; \frac{9}{10}; \frac{19}{20}; \frac{37}{40}$
18. $\frac{7}{9}; \frac{5}{18}; \frac{25}{27}; \frac{2}{3}$
19. $\frac{1}{2}; \frac{13}{22}; \frac{5}{11}; \frac{3}{4}$
20. $\frac{3}{7}; \frac{5}{21}; \frac{19}{42}; \frac{1}{2}$

Exercise 24

Work answers for:

1. $1\frac{1}{2} \times 2$
2. $3\frac{1}{4} \times 8$
3. $5 \times 1\frac{1}{5}$

4. $2 \times 3\frac{1}{2}$
5. $6\frac{1}{4} \times 2$
6. $7 \times 3\frac{1}{2}$
7. $5 \times 5\frac{1}{2}$
8. $\frac{1}{2} \times 3\frac{1}{4}$
9. $\frac{1}{4} \times 7$
10. $1\frac{1}{2} \times 10$
11. $7\frac{1}{2} \times 20$
12. $10 \times 3\frac{1}{4}$
13. $5\frac{1}{4} \times 12$
14. $1\frac{1}{2} \times 1\frac{1}{3}$
15. $\frac{1}{2} \times \frac{2}{3} \times \frac{5}{6}$
16. $\frac{1}{8} \times 3\frac{1}{2}$
17. $\frac{2}{3} \times \frac{3}{4}$
18. $\frac{7}{9} \times \frac{1}{14}$
19. $\frac{3}{5} \times \frac{5}{9}$
20. $1\frac{1}{2} \times 1\frac{1}{2} \times 1\frac{1}{2}$
21. $\frac{3}{4} \times 2$
22. $2\frac{1}{2} \times 8$
23. $1\frac{1}{2} \times 10$
24. $\frac{21}{8} \times 4$
25. $1\frac{1}{2} \times 18$
26. $3\frac{1}{2} \times 20$
27. $\frac{1}{10} \times 2$
28. $5\frac{1}{2} \times 12$
29. $3\frac{1}{8} \times 6$
30. $5\frac{1}{3} \times 3$

Exercise 25

Work answers for:

1. $\frac{1}{2} - \frac{1}{4}$
2. $\frac{1}{3} + \frac{2}{3}$
3. $1\frac{1}{8} - \frac{5}{8}$
4. $1\frac{1}{3} - \frac{5}{6}$
5. $\frac{5}{6} - \frac{1}{3}$
6. $\frac{1}{3} - \frac{1}{4}$
7. $1\frac{1}{8} - \frac{3}{4}$
8. $1\frac{1}{2} - \frac{1}{3}$
9. $4 - \frac{1}{4}$
10. $5 - \frac{1}{5}$
11. $\frac{2}{3} - \frac{1}{4}$
12. $1\frac{1}{8} - \frac{3}{16}$
13. $\frac{1}{4} - \frac{1}{12}$
14. $2\frac{1}{6} - 1$
15. $1 - \frac{1}{6}$
16. $1\frac{1}{2} - \frac{1}{4}$
17. $\frac{1}{8} - \frac{1}{12}$
18. $\frac{1}{2} - \frac{1}{12}$
19. $\frac{2}{3} - \frac{2}{5}$.
20. $2 - \frac{1}{4}$.

Exercise 26

Work answers for the following remembering to state the **unit** in each case.

1. $\frac{1}{5}$ of £20.
2. $\frac{1}{3}$ of £90.
3. $\frac{1}{4}$ of 80 pence.
4. $\frac{2}{3}$ of £60.
5. $\frac{1}{8}$ of £1.
6. $\frac{1}{2}$ of £1·50.
7. $\frac{1}{3}$. of £2·40.
8. $\frac{1}{8}$ of £64.
9. $\frac{1}{10}$ of £45.
10. $\frac{2}{3}$ of £9.
11. $\frac{1}{5}$ of £100.
12. $\frac{5}{6}$ of £18.
13. $\frac{1}{4}$ of £18.
14. $\frac{1}{10}$ of £110.
15. $\frac{3}{10}$ of £45.
16. $\frac{1}{8}$ of £120.
17. $\frac{1}{3}$ of £2·70.
18. $\frac{3}{8}$ of £160.
19. $\frac{1}{7}$ of £49·70.
20. $\frac{1}{4}$ of £60.

Exercise 27

Give the following fractions as percentages.

1. $\frac{1}{2}$
2. $\frac{1}{4}$
3. $\frac{3}{4}$
4. $\frac{1}{5}$

5. $\frac{2}{5}$ 6. $\frac{3}{5}$ 7. $\frac{1}{10}$ 8. $\frac{3}{10}$
9. $\frac{7}{10}$ 10. $\frac{9}{10}$ 11. $\frac{1}{20}$ 12. $\frac{3}{20}$
13. $\frac{7}{20}$ 14. $\frac{9}{20}$ 15. $\frac{11}{20}$ 16. $\frac{13}{20}$
17. $\frac{17}{20}$ 18. $\frac{19}{20}$ 19. $\frac{1}{50}$ 20. $\frac{3}{20}$

Exercise 28

Give the following percentages as fractions in their lowest terms.

1. 3%   2. 4%   3. 17%   4. 18%
5. 20%   6. 10%   7. 8%   8. 28%
9. 40%   10. 45%   11. 55%   12. 32%
13. 48%   14. 80%   15. 75%   16. 92%
17. 35%   18. 29%   19. 36%   20. 41%

Exercise 29

Express as percentages:

1. $2\frac{1}{2}$ 2. $7\frac{1}{4}$ 3. $9\frac{3}{4}$ 4. $1\frac{1}{10}$
5. $3\frac{3}{10}$ 6. $2\frac{3}{4}$ 7. $5\frac{3}{4}$ 8. $8\frac{1}{4}$
9. $2\frac{7}{10}$ 10. $8\frac{9}{10}$ 11. $1\frac{1}{20}$ 12. $5\frac{3}{20}$
13. $7\frac{1}{20}$ 14. $14\frac{1}{10}$ 15. $1\frac{1}{20}$ 16. $5\frac{1}{50}$
17. $5\frac{1}{2}$ 18. $3\frac{9}{10}$ 19. $14\frac{17}{20}$ 20. $15\frac{9}{20}$

Exercise 30

Express as fractions in lowest terms:

1. 250%   2. 130%   3. 80%   4. 110%
5. 900%   6. 750%   7. 118%   8. 220%
9. 340%   10. 120%   11. 610%   12. 380%
13. 425%   14. 111%   15. 218%   16. 692%
17. 138%   18. 1416%   19. 515%   20. 275%

Exercise 31

Express as percentages:

1. 1·3   2. 3·25   3. 0·04   4. 0·18

5.	2·74	6.	1·21	7.	0·08	8.	1·23
9.	1·18	10.	4·46	11.	2·2	12.	10·4
13.	0·01	14.	0·1	15.	7·14	16.	3·41
17.	1·214	18.	2·684	19.	1·3	20.	1·003

Exercise 32

Express as decimal fractions:

1.	120%	2.	87%	3.	41%	4.	226%
5.	83%	6.	92%	7.	134%	8.	189%
9.	114%	10.	2%	11.	8%	12.	39%
13.	4·8%	14.	5·64%	15.	91%	16.	324%
17.	38%	18.	12%	19.	1%	20.	2000%

Exercise 33

Work answers to the following:

1. 2% of £200.
2. 8% of £4.
3. 20% of £300.
4. 110% of £1.
5. 120% of £40.
6. 80% of £100.
7. 7% of £50.
8. 5% of £40.
9. 90% of £1.
10. 3% of £20.
11. 15% of £4.
12. 150% of £100.
13. 160% of £6.
14. 1% of £1000.
15. 1% of £1 000 000.
16. 75% of £2.
17. 25% of £40.
18. 12% of £500.
19. 40% of £1100.
20. 50% of £2·50.

Exercise 34

Work answers for:

1. Add $\frac{1}{5}$ of £2 to 2% of £100.
2. Take $\frac{1}{8}$ of £16 from £4·50.
3. Add $\frac{1}{4}$ of £1 to 3% of £300.
4. Add $\frac{1}{5}$ of £8 to $\frac{1}{4}$ of £12.
5. Take $\frac{3}{4}$ of £100 frm £150.
6. Add $\frac{1}{8}$ of £14 to $\frac{1}{4}$ of £6.
7. Take £110·50 from 4 × £49·50.
8. Divide 20% of £100 by 4.
9. Multiply 30% of £200 by 10.
10. Divide 3% of 1000 by 4.
11. Add 40% of £1000 to £750.
12. Take 70% of £400 from £324·50.
13. Add $\frac{1}{8}$ of £184 to $\frac{1}{6}$ of £12.
14. Take 4 × £93 from 10% of £1 000 000.

15. Add $\frac{7}{10}$ of £50 to $\frac{3}{10}$ of £300.
16. Take £1·30 from 4% of £50.
17. Add 80% of £800 to 20% of £600.
18. Take £1·39 from 40% of £80.
19. Add $\frac{1}{8}$ of £4 to $\frac{1}{5}$ of £3.
20. Subtract $\frac{3}{10}$ of £40 from £16·82.

Exercise 35

Express in centimetres:

1. 1·2 m
2. 1 km
3. 14 m
3. 1·8 m
5. 1·5 km
6. 2·8 km
7. 0·08 km
8. 1·84 m
9. 3·72 m
10. 15 m
11. 0·1 m
12. 0·01 km
13. 2·01 km
14. 0·001 m
15. 0·08 m
16. 0·0001 km
17. 8 mm
18. 84 mm
19. 8623 mm
20. 12846 mm

Exercise 36

Express in metres:

1. 8 km
2. 0·01 mm
3. 123 mm
4. 1·84 km
5. 7·2 km
6. 85 mm
7. 1982 mm
8. 1·46 km
9. 0·041 km
10. 2·01 km
11. 1 000 000 mm
12. 8 mm
13. 8·86 km
14. 84·2 km
15. 382 cm
16. 9214 cm
17. 0·001 cm
18. 8·6 cm
19. 12·8 km
20. 8884 cm

Exercise 37

Work answers to the following. In each case give your answer in metres.

1. 2·3 m + 1 km.
2. 1·82 km + 3 m.
3. 2·4 m + 1·3 cm.
4. 18 cm + 186 cm + 17·8 m.
5. 1·2 m + 14 cm + 140 mm.
6. 8·4 km + 8·2 m.
7. 862 m + 32 km.
8. 428 mm + 42 m + 880 cm.
9. 3 km + 30 m + 300 cm.
10. 23 km + 38 cm + 840 m.
11. 3 m + 2·8 m + 6·08 m.
12. 800 cm + 8000 mm.
13. 1·2 km + 1·8 km + 1·3 km.
14. 1·74 km + 1·21 km + 300 cm.
15. 850 cm + 8500 mm + 8 m.
16. 9·2 km + 1·3 km + 0·8 km.
17. 2400 mm + 1800 cm.
18. 1800 mm + 1840 mm + 1600 cm.
19. 1·01 km + 1·02 km + 1·03 km.
20. 84 m + 87 m + 82 000 cm.

Exercise 38

Work the following areas. The measurements given in each case are for a rectangle. Give your answers in **square metres**.

1.	Length 18 m	Breadth 9 m
2.	Length 2 m	Breadth 0·2 m
3.	Length 1 m	Breadth 0·5 m
4.	Length 50 cm	Breadth 20 cm
5.	Length 60 cm	Breadth 50 cm
6.	Length 80 cm	Breadth 40 cm
7.	Length 12 m	Breadth 1000 cm
8.	Length 10 m	Breadth 100 000 mm
9.	Length $1\frac{1}{2}$ m	Breadth $1\frac{1}{4}$ m
10.	Length 18 m	Breadth 10 m
11.	Length 1·8 m	Breadth 1·2 m
12.	Length 100 cm	Breadth 80 m
13.	Length 1 m	Breadth 60 cm
14.	Length 8 m	Breadth 600 cm
15.	Length 6 m	Breadth 4 m
16.	Length 4 m	Breadth 300 cm
17.	Length 2 m	Breadth 1·5 m
18.	Length $\frac{3}{4}$ m	Breadth $\frac{1}{8}$ m
19.	Length $\frac{1}{4}$ m	Breadth $\frac{1}{6}$ m
20.	Length 10 m	Breadth 900 cm

Exercise 39

What distance is covered, in kilometres, if a man travels?

1. For 4 hours at 20 km/hour.
2. For 6 hours at 8 km/hour.
3. For $2\frac{1}{2}$ hours at 80 km/hour.
4. For 1 hour at 60 km/hour.
5. For 10 hours at 35 km/hour.
6. For $1\frac{1}{2}$ hours at 4 km/hour.
7. For $2\frac{1}{4}$ hours at 8 km/hour.
8. For 12 hours at 8·5 km/hour.
9. For 8 hours at 90 km/hour.
10. For $3\frac{1}{2}$ hours at 70 km/hour.
11. For $7\frac{1}{2}$ hours at 10 km/hour.
12. For 30 minutes at 8 km/hour.
13. For 20 minutes at 50 km/hour.
14. For 40 minutes at 100 km/hour.
15. For 90 minutes at 70 km/hour.
16. For 10 minutes at 120 km/hour.
17. For 5 minutes at 144 km/hour.
18. For $5\frac{1}{2}$ hours at 20 km/hour.
19. For $1\frac{1}{10}$ hours at 100 km/hour.
20. For 14 hours at 30 km/hour.

Exercise 40

At what speed, in km/hour will the following journeys be done?

1. 20 km in 2 hours.
2. 100 km in 5 hours.
3. 250 km in 10 hours.
4. 14 km in 1 hour.
5. 100 km in $2\frac{1}{2}$ hours.
6. 210 km in 3 hours.
7. 80 km in 8 hours.
8. 240 km in 3 hours.
9. 50 km in 2 hours.
10. 150 km in $1\frac{1}{2}$ hours.
11. 100 km in 15 minutes.
12. 200 km in 40 minutes.
13. 80 km in 4 hours.
14. 120 km in 15 hours.
15. 225 km in $7\frac{1}{2}$ hours.
16. 90 km in 2 hours.
17. 10 km in 1 minute.
18. 20 km in 10 minutes.
19. 1800 km in 90 minutes.
20. 1 km in 30 seconds.

Exercise 41

Find the time taken for the following journeys giving your answers in hours.

1. 150 km at 50 km/hour.
2. 300 km at 10 km/hour.
3. 1000 km at 200 km/hour.
4. 100 km at 40 km/hour.
5. 350 km at 70 km/hour.
6. 80 km at 4 km/hour.
7. 800 km at 100 km/hour.
8. 90 km at 45 km/hour.
9. 140 km at 40 km/hour.
10. 20 km at $2\frac{1}{2}$ km/hour.
11. 600 km at 100 km/hour.
12. 450 km at 30 km/hour.
13. 96 km at 16 km/hour.
14. 78 km at 39 km/hour.
15. 160 km at 52 km/hour.
16. 180 km at 45 km/hour.
17. 3240 km at 162 km/hour.
18. 1890 km at 90 km/hour.
19. 20 km at $\frac{1}{2}$ km/hour.
20. 100 km at 24 km/hour.

A Basic Mathematics Workbook

Exercise 42

From the following table, find, in each case, the gain or loss on the sale of 10 articles. The articles being bought and sold are of the same kind.

	Cost price of 1 article	Selling price of 5 articles
1.	£1	£6
2.	50p	£1·80
3.	£10	£55
4.	£8·50	£60
5.	£0·40	£2·50
6.	£1·20	£5
7.	£1·10	£4·50
8.	£3·20	£15
9.	£2·25	£15
10.	£3	£10
11.	£1·80	£7·50
12.	£8·80	£50
13.	£6·20	£25·50
14.	£0·20	£0·50
15.	£7·40	£32·50
16.	£5	£22·50
17.	£1·06	£5
18.	£0·90	£4
19.	£108	£500
20.	£2·40	£10

Arithmetic

Exercise 43

Work the volume of each of the following cuboids. Give your answer in cm³ in each case.

1. $l = 20$ cm; $b = 10$ cm; $h = 1$ cm.
2. $l = 10$ cm; $b = 5$ cm; $h = 1$ cm.
3. $l = 8$ cm; $b = 2$ cm; $h = 1$ cm.
4. $l = 1$ m; $b = 50$ cm; $h = 10$ cm.
5. $l = \frac{1}{2}$ m; $b = 40$ cm; $h = 30$ cm.
6. $l = \frac{1}{4}$ m; $b = 18$ cm; $h = 10$ m.
7. $l = 1\frac{1}{2}$ m; $b = 100$ cm; $h = 50$ cm.
8. $l = \frac{3}{4}$ m; $b = 40$ cm; $h = 10$ cm.
9. $l = 8\frac{1}{2}$ cm; $b = 4$ cm; $h = 2$ cm.
10. $l = 10\frac{1}{2}$ cm; $b = 8$ cm; $h = 6$ cm.
11. $l = 2\frac{1}{2}$ cm; $b = 2$ cm; $h = 1\frac{1}{2}$ cm.
12. $l = 1$ cm; $b = \frac{1}{2}$ cm; $h = \frac{1}{4}$ cm.
13. $l = 80$ cm; $b = 40$ cm; $h = 20$ cm.
14. $l = 2\frac{1}{2}$ m; $b = 1\frac{1}{4}$ m; $h = 100$ cm.
15. $l = 10$ cm; $b = 8$ cm; $h = \frac{1}{4}$ cm.
16. $l = 24$ cm; $b = 12$ cm; $h = 10$ cm.
17. $l = 1$ m; $b = \frac{3}{4}$ m; $h = \frac{1}{2}$ m.
18. $l = 20$ cm; $b = 10$ cm; $h = 6$ cm.
19. $l = 2\frac{1}{8}$ m; $b = 80$ cm; $h = 60$ cm.
20. $l = \frac{1}{4}$ m; $b = \frac{1}{4}$ m; $h = \frac{1}{10}$ m.

Exercise 44

In each of the questions 1 to 20, work out how many litres of water each of the cuboids in **Exercise 43** would hold, if the above measurements were **inside** measurements.

Exercise 45

Give the following speeds in metres per second.

1. 20 km/hour.
2. 10 km/minute.
3. 100 km/hour.
4. 5 km/minute.
5. 2 km/second.
6. 8 km/minute.
7. 80 m/minute.
8. 10 m/minute.
9. 8 km/hour.
10. 80 km/hour.
11. 6 cm/second.
12. 120 cm/second.

18 A Basic Mathematics Workbook

13. 24 m/minute.
14. 86 km/hour.
15. 240 cm/second.
16. 460 cm/minute.
17. 1 km/minute.
18. 400 km/hour.
19. $1\frac{1}{2}$ km/second.
20. $2\frac{1}{4}$ km/second.

Exercise 46

1. How many steps of 80 cm each are taken in walking 4 km?
2. How many lengths each 30 cm long can be cut from 2 metres? What length remains?
3. Multiply 24 cm by 8 and take your result from 4 metres. Answer in centimetres.
4. What length is travelled in 20 seconds at a rate of 0·4 km/minute?
5. How many times can 0·2 litres be taken from 16 litres of water?
6. What is the capacity, in litres, of a cuboid with internal measurements of $l = 40$ cm, $b = 30$ cm and $h = 20$ cm?
7. How many steps of 100 cm each are taken in walking 21 km?
8. Add 1·4 m, 32 cm and 484 mm. Give your answer in metres.
9. How many millimetres are in 34·2 cm?
10. How many cups can be filled from a 4·8 litre can of water if each cup holds 24 cm³ of water?
11. Multiply 8·4 cm by 6 and subtract your answer from 1 metre. Answer in metres.
12. What is the capacity, in litres, of a cuboid with internal measurements of: $l = \frac{1}{2}$ m, $b = 20$ cm, $h = 10$ cm?
13. How many lengths of string, each of 1·5 m, could be cut from a ball of string containing 0·45 km?
14. Add 1·3 m and 89 cm and take the result from 4·2 m. Answer in centimetres.
15. Subtract 1·2 cm from 6·4 cm and add your result to 0·02 metres. Answer in centimetres.
16. From 14·2 cm take 89 mm. Add your result to 0·82 metres. Answer in centimetres.
17. Multiply 22 mm by 40. Subtract your answer from 1 metre. Answer in centimetres.
18. How many centimetres are in 284 km?
19. Express as fractions of 1 litre: (a) 44 cm³; (b) 880 mm³; (c) 8·2 cm³.
20. Express (a), (b) and (c) of question 19. as percentages.

Exercise 47

Perform the following additions. In each case, give your answer in grammes.

1. 2·4 kg + 0·01 kg.
2. 32 g + 0·04 kg.
3. 484 mg + 382 g.
4. 1·23 kg + 0·028 kg.

5. 89 mg + 792 mg.
6. 8840 g + 0·02 kg.
7. 22 mg + 38 mg + 8 g.
8. 0·8 g + 44 mg.
9. 22 kg + 84 g + 760 mg.
10. 22 400 mg + 88 600 g.
11. 3·12 kg + 0·071 kg.
12. 4840 mg + 1000 mg.
13. 27 kg + 2·7 kg + 0·27 kg.
14. 44 mg + 4400 mg.
15. 8806 g + 8·806 g.
16. 0·021 kg + 0·04 kg.
17. 5500 g + 555 mg.
18. 0·03 kg + 3300 g.
19. 2984 g + 7632 mg.
20. 3·14 kg + 80 mg + 800 g.

Exercise 48

Perform the following subtractions. In each case, give your answer in grammes.

1. From 2·4 g take 84 mg.
2. Take 23 g from 1·2 kg.
3. Take 894 mg from 1246 g.
4. Take 2·4 kg from 8·62 kg.
5. From 8 kg take 832 g.
6. Take 7 g from 8·2 g.
7. Take 23 g from 0·43 kg.
8. From 84 g take 481 mg.
9. Take 31 g from 0·64 kg.
10. From 1·02 kg take 834 g.

Exercise 49

Work answers to the following questions.

1. How many mg are in 3·4 kg?
2. How many packs each weighing 3·2 kg could be weighed off from 1 tonne of sugar? What weight of sugar, if any, would be left over?
3. Express 32 g as a fraction of 2 kg.
4. Express 8 g as a percentage of 0·5 kg.
5. Find the total weight of 42 parcels each weighing 838 g. Answer in kg.
6. Express 84 321 g in kg.
7. Express the sum of 24 g and 49 g as a decimal fraction of 1 kg.
8. How many mg are in 2·86 kg?
9. Express 84 g as a percentage of 1 kg.
10. Find the total weight of a million packages each weighing 1·2 g. Answer in tonnes.

Exercise 50

In each case work out the **extra** paid by buying on hire purchase.

	Cash Price of Article	Deposit	Amount of Each Payment	Number of Payments
1.	£450	£45	£20	24
2.	£80	£20	£5·80	12
3.	£220	£28	£18·20	12
4.	£24	£4	£2·10	12
5.	£1020	£100	£27·20	36
6.	£38	£5	£1·50	24
7.	£66·50	£6	£6·20	12
8.	£1026	£85	£41·20	24
9.	£96·50	£32	£7·10	12
10.	£642	£200	£13·40	36
11.	£27·50	£3	£2·80	12
12.	£82	£10	£7·05	12
13.	£46·50	£4	£2·10	24
14.	£19	£1	£1·82	12
15.	£184	£31	£15·08	12
16.	£25·50	£8	£1·86	12
17.	£250	£84	£8·32	24
18.	£4000	£1400	£81·10	36
19.	£364·50	£120	£12·84	24

Exercise 51

In each case give the **selling price** to make a 10% gain on the cost price shown. Round **up** to nearest 1p where necessary.

1. C.P. = £4.
2. C.P. = £6.
3. C.P. = £1
4. C.P. = £406.
5. C.P. = £22·50.
6. C.P. = £17·45.
7. C.P. = £84·84.
8. C.P. = £120.
9. C.P. = £24·90.
10. C.P. = £12·27.
11. C.P. = £46.
12. C.P. = 80p.
13. C.P. = £11·50.
14. C.P. = 40p.
15. C.P. = 95p.
16. C.P. = £4·40.
17. C.P. = £33·33.
18. C.P. = £1000.
19. C.P. = £727·50.
20. C.P. = 17p.

Exercise 52

Use the cost prices given in **Exercise 51** to work out what the selling prices must have been to give a **10% loss** in each case. Round **down** to nearest 1p where necessary.

A Basic Mathematics Workbook

Exercise 53

Find the **interest** due on the money borrowed in each of the questions 1 to 20 shown in the table below. Use the **simple interest** formula

$$I = \frac{P \times R \times T}{100}$$

Round up to nearest 1p where necessary.

	Sum of Money Borrowed	Rate of Interest Per Year	Duration of Borrowing
1.	£400	12%	6 months
2.	£100	8%	1 year
3.	£50	19%	1 year
4.	£1000	20%	3 months
5.	£550	10%	3 years
6.	£220	14%	9 months
7.	£660	9%	6 years
8.	£320	11%	8 months
9.	£3000	18%	1 month
10.	£290	10%	1 year
11.	£750	8%	2 years
12.	£85	18%	2 months
13.	£294	15%	6 months
14.	£10	10%	2 years
15.	£94	12%	3 months
16.	£1050	5%	2 years
17.	£74	10%	1 year
18.	£7500	15%	6 months
19.	£1200	12%	9 months
20.	£80 000	10%	2 years

Exercise 54

In each of the following questions 1 to 20 calculate **gross** wage earned. In each case, overtime is earned at **double time**.

	Basic Rate Per Hour	Basic Working Week	Actual Time worked in week	Gross Wage
1.	£1·40	32 hours	36 hours	
2.	£2·10	36 hours	40 hours	
3.	£1·08	41 hours	47 hours	
4.	£3·50	28 hours	32 hours	
5.	£2·70	34 hours	36 hours	
6.	£1·25	42 hours	48 hours	
7.	£2·40	37 hours	37 hours	
8.	£2·60	36 hours	38 hours	
9.	£1·90	38 hours	39 hours	
10.	£5	31 hours	33 hours	
11.	£1·20	42 hours	50 hours	
12.	£1·75	37 hours	43 hours	
13.	£2·24	38 hours	42 hours	
14.	£4·40	38 hours	38 hours	
15.	£1·92	42 hours	52 hours	
16.	£3·20	32 hours	36 hours	
17.	£1·62	38 hours	48 hours	
18.	£2	32 hours	42 hours	
19.	£2·50	38 hours	38 hours	
20.	£3·15	42 hours	44 hours	

SECTION 2

ALGEBRA

Exercise 1

Work number answers for 1 to 20 if $t = 3$, $p = 4$ and $x = 5$.

1. $t + p$
2. $5 + p$
3. $x + p$
4. $3(t + x)$
5. $7 - p$
6. $p + 3x$
7. $4(x + p)$
8. $\dfrac{x}{5}$
9. $2p$
10. $\dfrac{1}{9}(p + x)$
11. $4(x - p)$
12. $xp + t$
13. $pt + x$
14. ptx
15. xpt
16. xtp
17. $x + p + t$
18. $xp + pt + xt$
19. $3x + 2p$
20. $4t - x$

Exercise 2

Using the same values as in **Exercise 1**, work number answers for 1 to 20.

1. $3x + 2p$
2. $3p + 4x$
3. $21 + 7p$
4. $3t + 4x$
5. $5p - x$
6. $3t - 1$
7. $3(x + p - t)$
8. $4x + 7p$
9. $5x - 2p + 3t$
10. $2x - t$
11. $7x + 3p$
12. $5x + 2p - t$
13. $4x$
14. $11p$
15. $17t$
16. $\dfrac{x + 3p}{7}$
17. $\dfrac{3x - p + t}{7}$
18. $\dfrac{x + 8t + 7p}{19}$
19. $24t - p$
20. $3(2x + 4p + 7t)$

Exercise 3

If $a = 1$, $b = 3$ and $c = 2$ work the number answers for:

1. a^2
2. b^2
3. c^2
4. $(ab)^2$
5. $(bc)^2$
6. $a^2 + b^2$
7. $c^2 + b^2$
8. a^3
9. b^3

10. c^3
11. $(ac)^3$
12. $3a^2$
13. $3(a^2 + b^2)$
14. $\dfrac{a^2 + c^2}{5}$
15. $\dfrac{b^3 + c^3}{7}$
16. $\dfrac{a^3 + b^2}{10}$
17. $\dfrac{bc}{6}$
18. $(2a)^2$
19. $(3c)^2$
20. $(3ab)^3$

Exercise 4

Give simpler answers, where possible:

1. $2a + 3a - a$.
2. $5b + 2b - a$.
3. $6a + 2a + 8a$.
4. $5b - c$.
5. $3a + 5x + 7x - a$.
6. $3c + 2c + 8c + 1$.
7. $4p + 2q + 6p - q$.
8. $3a + 6b + 2c$.
9. $5x + 2x + 3y - x + 4y$.
10. $5t + 7t - 3t$.
11. $6a + 2a + 4b + c$.
12. $2c + c + c + 1$.
13. $6t + 5t - t + 4t$.
14. $a + 2c + 3d$.
15. $2d - d + 3d - d$.
16. $7d + 2d - 8d + 1$.
17. $14t + 6t - 19t$.
18. $c + 3c + 4c$.
19. $c + c + c + c + d$.
20. $2t + 3p - 4$.

Exercise 5

If $a = 2, b = 1, c = 3, x = 5, p = 4, q = 6, y = 7, t = 11, d = 9$, give your answers for 1 to 20 in **Exercise 4 as numbers.**

Exercise 6

Write in a simpler way:

1. $3a \times 2b$.
2. $5a^2 \times 2a$.
3. $4b \times 8b$.
4. $x \times 2y \times 3x$.
5. $5ab \times 2c$.
6. $ab \times 2a$.
7. $3c^2 \times 2d$.
8. $abc \times 2a$.
9. $mn \times 1mn$.
10. $p^2q \times 2p$.
11. $3qt \times 2t$.
12. $3t \times 2t^2$.
13. $5ab \times 5ab \times 5ab$.
14. $2m \times 3m \times 4m$.

26 A Basic Mathematics Workbook

15. $3 \times 6a \times 2a^2$.

16. $2p^2q \times 8$.

17. $4a^2b \times 2$.

18. $6p^2q \times 4$.

19. $3mn \times 2mt$.

20. $5ad \times 2ac$.

Exercise 7

Write in a simpler way:

1. $\dfrac{36ab}{3}$

2. $\dfrac{24p^2}{12}$

3. $\dfrac{48a^2}{16a}$

4. $\dfrac{8mn}{2m}$

5. $\dfrac{12a^2c}{4ca^2}$

6. $\dfrac{2abc}{4b}$

7. $\dfrac{5pq}{10q}$

8. $\dfrac{12m^2}{2m}$

9. $\dfrac{16a^2b}{8b}$

12. $\dfrac{20p^2q}{10}$

11. $\dfrac{14a^2b}{7ab}$

12. $\dfrac{2p}{4}$

13. $\dfrac{8q}{16}$

14. $\dfrac{48q^4}{16}$

15. $\dfrac{8abc}{bc}$

16. $\dfrac{12m^4}{3}$

17. $\dfrac{abd}{2bd}$

18. $\dfrac{16}{2a}$

19. $\dfrac{14a^2b}{7b}$

20. $\dfrac{12m^2n^3}{m^2n}$

Exercise 8

In each case write down the cost of x books, in £'s, if:

1. p books cost t pence, in total.
2. p books cost £3, in total.
3. 3 books cost £d, in total.
4. One book costs c pence.
5. 12 books cost £36, in total.
6. y books cost £e, in total.
7. One book costs ad pence.
8. q books cost £t, in total.
9. t books cost £m, in total.
10. 1 book costs mn in pence.

Exercise 9

In each case write down, in pence, the cost of 1 item, if:

1. 3 items cost a total of £x.
2. x items cost a total of £y.
3. c items cost ad pence, altogether.
4. d items cost £pq, in all.
5. 7 items cost £14t, altogether.
6. x items cost a total of £a.

7. *m* items cost *n* pence, in all.
8. *ad* items cost £*x*, altogether.
9. 10*p* items cost a total of *y* pence.
10. *t* items cost a total of £*xy*.
11. 2*t* items cost *ac* pence, in total.
12. 3*q* items cost a total of £9.
13. *ma* items cost *t* pence altogether.
14. 5 items cost £15 in total.
15. 3 items cost £15*x* in total.
16. *n* items cost *m* pence altogether.
17. 1 item cost £*x*.
18. 1 item cost £3*y*.
19. 4*t* items cost a total of £4*x*.
20. *d* items cost £*d* altogether.

Exercise 10

Express in metres, giving your answer in simplest form:

1. 2*x* cm
2. 3*t* mm
3. *m* km
4. 3*t* cm
5. 15*a* mm
6. 2*t* km
7. $\frac{t}{a}$ mm
8. $\frac{2t}{m}$ cm
9. $\frac{3t}{2}$ km
10. 4*xy* km
11. $\frac{3x}{4}$ mm
12. 12*ac* km
13. $\frac{a}{c}$ mm
14. $\frac{2a}{d}$ km
15. *at* mm
16. *pq* km
17. 3*p* mm
18. 12*q* cm
19. $\frac{3m}{2t}$ cm
20. 12*ad* mm

Exercise 11

Express in grammes, giving your answer in simplest form:

1. 3*t* kg
2. *ar* mg
3. $\frac{p}{q}$ kg
4. (*a* + *b*) kg
5. $\frac{2t}{5}$ mg
6. $\frac{5a}{c}$ mg
7. 2(*a* + *d*) mg
8. 5(2*d* + 3) kg
9. $\frac{3m}{n}$ kg
10. $\frac{5(a+c)}{2}$ mg
11. *abc* mg
12. 2*cd* mg
13. 5*f* mg
14. 144*a* kg
15. $\frac{t}{p}$ mg

16. $\dfrac{2m}{5n}$ mg **17.** $\dfrac{24a}{d}$ kg **18.** $2t$ kg

19. $17pq$ mg **20.** $\dfrac{pq}{ad}$ mg

Exercise 12

Work answers for:

1. c metres $+ ad$ cm; answer in metres.
2. ab cm $+ d$ mm; answer in mm.
3. $2m$ cm $+ 3n$ cm; answer in cm.
4. $\dfrac{p}{q}$ metres $+ d$ cm; answer in cm.
5. $2a$ cm $+ 3d$ mm; answer in cm.
6. $5p$ metres $+ 3q$ cm; answer in cm.
7. $\dfrac{a}{d}$ metres $+ \dfrac{d}{a}$ cm; answer in metres.
8. $(2t + 3p)$ metres $+ d$ cm; answer in cm.
9. mn mm $+ 3cd$ cm; answer in mm.
10. $2ac$ cm $+ 3pq$ mm; answer in mm.

Exercise 13

We will use the formula $c = 2d + t$ throughout this exercise. In 1 to 20, find c if:

1. $d=4, t=2\tfrac{1}{2}$.
2. $d=1, t=3\tfrac{1}{2}$.
3. $d=5, t=1$.
4. $d=2, t=4$.
5. $d=1\tfrac{1}{2}, t=10$.
6. $d=4, t=3\tfrac{1}{2}$.
7. $d=15, t=2$.
8. $d=100, t=4$.
9. $d=10, t=44$.
10. $d=39, t=12$.
11. $d=15\tfrac{1}{2}, t=1$.
12. $d=42, t=4$.
13. $d=4, t=32$.
14. $d=7, t=13$.
15. $d=16, t=4$.
16. $d=4, t=16$.
17. $d=9, t=8$.
18. $d=4, t=19$.
19. $d=3\tfrac{1}{2}, t=10$.
20. $d=10\tfrac{1}{2}, t=4$.

Exercise 14

We will use the formula $g = \tfrac{1}{2}(a + 2p)$ throughout this exercise. In 1 to 20, find g if:

1. $a=4, p=1$.
2. $a=3, p=\tfrac{1}{2}$.
3. $a=6, p=3$.
4. $a=7, p=\tfrac{1}{2}$.
5. $a=17, p=\tfrac{1}{2}$.
6. $a=4, p=7$.

7. $a = 3, p = 3\frac{1}{2}$.
8. $a = 2, p = 17$.
9. $a = 5, p = 2\frac{1}{2}$.
10. $a = 17, p = 5\frac{1}{2}$.
11. $a = 2, p = 7$.
12. $a = 6, p = 14$.
13. $a = 10, p = 3$.
14. $a = 11, p = 3\frac{1}{2}$.
15. $a = 40, p = 2$.
16. $a = 42, p = 6$.
17. $a = 32, p = 4$.
18. $a = 24, p = 2$.
19. $a = 18, p = 13$.
20. $a = 15, p = \frac{1}{2}$.

Exercise 15

We will use the formula $T = 2(t - q)$ throughout this exercise. In 1 to 20, find T if:

1. $t = 7, q = 1$.
2. $t = 6, q = 2$.
3. $t = 14, q = 1$.
4. $t = 11, q = 1$.
5. $t = 20, q = 2$.
6. $t = 41, q = 7$.
7. $t = 3, q = 2$.
8. $t = 21, q = 17$.
9. $t = 14, q = 11$.
10. $t = 19, q = 11$.
11. $t = 22, q = 14$.
12. $t = 18, q = 2$.
13. $t = 6, q = 5$.
14. $t = 10, q = 7$.
15. $t = 12, q = 9$.
16. $t = 32, q = 31$.
17. $t = 27, q = 18$.
18. $t = 25, q = 23$.
19. $t = 78, q = 6$.
20. $t = 102, q = 101$.

Exercise 16

Write in simplest form:

1. $3a + 21a$.
2. $14c - 7c$.
3. $21x + 3x$.
4. $12xy + 2xy$.
5. $3ac - 2ac$.
6. $5p - 2p$.
7. $2a + 12a$.
8. $17p - 3p$.
9. $1q + 8q$.
10. $2a + 5a - a$.
11. $3a + 4a + 6a$.
12. $18a - 17a$.
13. $2at + 5at$.
14. $3am + 2am$
15. $7ac - ac$.
16. $x + 3x + 2x$.
17. $5x - x - x$.
18. $3x + 2x - x$.
19. $2t + 3t + 5t$.
20. $17x - 14x - 2x$.

Exercise 17

Write in simplest form:

1. $3a + 2a + c$.
2. $2p - p + q$.
3. $5q + 3q$.
4. $17a + 2a + 5b$.
5. $4ab + 2ab - ab$.
6. $7p + 11p$.
7. $x + 2x + y$.
8. $3y + 2y - y + a$.
9. $2y - y$.
10. $17ab + 3ab - ab$.

11. $5t + 8t - 7t$.
12. $16t - 2t$.
13. $5ab + 11ab - 10ab$.
14. $14x + 3x - 12x$.
15. $2ab + 6ab - 3ab$.
16. $2x + 4x + 11y + 2y$.
17. $3x + 3x + 3x + 2y$.
18. $5a + 3a + 6b + b$.
19. $2p + 3p + 4p + 2q$.
20. $2c + 3c + d + 5d$.

Exercise 18

Write in simplest form:

1. $\dfrac{2a}{4}$
2. $3p \times 2q$
3. $4m \times 2n$.
4. $\dfrac{18pq}{27}$
5. $2t \times 7tp$.
6. $\dfrac{3am}{6m}$
7. $\dfrac{4tp}{8}$
8. $\dfrac{16ac}{24}$
9. $2x \times 3x \times 4$.
10. $5a \times 2a$.
11. $13ab \times 4$.
12. $\dfrac{27a^2 b}{9b}$
13. $3ab \times 4b$.
14. $2x \times 3x \times 4xy$.
15. $\dfrac{100a}{10}$
16. $\dfrac{12p^2}{4q}$
17. $\dfrac{3m^2}{6m}$
18. $4p \times 2p \times q$.
19. $\dfrac{30ab^2}{10b}$
20. $\dfrac{4m \times 10m}{20}$

Exercise 19

In numbers 1 to 20 say whether the statement is true or false.

1. $x + 5 = 6$ if $x = 1$.
2. $x - 2 = 4$ if $x = 6$.
3. $x + 7 = 8$ if $x = 2$.
4. $x - 11 = 9$ if $x = 20$.
5. $x + 3 = 15$ if $x = 11$.
6. $a + b = 8$ if $a = 2$ and $b = 6$.
7. $a - b = 4$ if $a = 5$ and $b = 1$.
8. $2 + c = 8$ if $c = 6$.
9. $c + d = 12$ if $c = 1$ and $d = 11$.
10. $c + d = 12$ if $c = 11$ and $d = 1$.
11. $c - 2 = 3$ if $c = 4$.
12. $c - 8 = 7$ if $c = 16$.
13. $a + 12 = 19$ if $a = 7$.
14. $a + m = n$ if $a = 2, b = 3, n = 5$.
15. $a - m = n$ if $a = 4, m = 1, n = 3$.
16. $2a = 18$ if $a = 11$.
17. $3p = 8$ if $p = 4$.
18. $12 + p = 15$ if $p = 3$.
19. $2 + t = 6$ if $t = 4$.
20. $t + q = m$ if $t = 1, q = 3$ and $m = 5$.

Exercise 20

Say whether the following statements are true or false.

1. $2a = 6$, if $a = 3$.
2. $5t = 15$, if $t = 3$.
3. $2x = 14$, if $x = 8$.
4. $5m = 10$, if $m = 2$.
5. $3n = 18$, if $n = 7$.
6. $5a = 25$, if $a = 5$.
7. $3p = 42$, if $p = 14$.
8. $2a = 32$, if $a = 18$.
9. $5t = 100$, if $t = 20$.
10. $17p = 34$, if $p = 2$.
11. $18m = 18$, if $m = 1\frac{1}{2}$.
12. $27m = 9$, if $m = \frac{1}{3}$.
13. $a + b = 19$, if $a = 11$ and $b = 8$.
14. $a - b = 2$, if $a = 3$ and $b = 1$.
15. $45x = 90$, if $x = 2$.
16. $4p + 1 = 17$, if $p = 4$.
17. $3m - 1 = 17$, if $m = 6$.
18. $2m + 5 = 7$, if $m = 2$.
19. $x + y = 2t$ if $x = 4$, $y = 6$ and $t = 5$.
20. $x + 1 = 11$, if $x = 7$.

Exercise 21

Find the value of x in each of the following equations.

1. $x + 2 = 3$.
2. $x + 1 = 5$.
3. $x + 7 = 9$.
4. $x - 1 = 11$.
5. $x + 7 = 8$.
6. $x + 17 = 19$.
7. $x + 5 = 17$.
8. $x + 3 = 11$
9. $x + 11 = 17$.
10. $x + 21 = 22$.
11. $x - 2 = 17$.
12. $x - 1 = 31$.
13. $x + 5 = 13$.
14. $x + 1 = 27$.
15. $x + 3 = 14$.
16. $x + 11 = 31$.
17. $x + 2 = 17$.
18. $x + 12 = 18$.
19. $x - 17 = 73$.
20. $x - 74 = 75$.

Exercise 22

Find the value of t in each of the following equations.

1. $t - 5 = 7$.
2. $t + 5 = 10$.
3. $t + 11 = 101$.
4. $t - 3 = 87$.
5. $t + 3 = 23$.
6. $t - 3 = 93$.
7. $t - 9 = 1$.
8. $t - 8 = 2$.
9. $t + 4 = 91$.
10. $t - 3 = 106$.
11. $t - 1 = 1000$.
12. $t + 15 = 16$.
13. $t - 19 = 27$.
14. $t + 56 = 57$.
15. $t - 3 = 206$.
16. $t + 17 = 32$.
17. $t - 3 = 75$.
18. $t - 5 = 46$.
19. $t - 11 = 12$.
20. $t + 1 = 79$.

32 A Basic Mathematics Workbook

Exercise 23

Find the value of c in each of the following equations.

1. $2c = 4$.
2. $3c = 81$.
3. $4c = 100$.
4. $5c = 15$.
5. $5c = 45$.
6. $15c = 60$.
7. $2c = 108$.
8. $3c = 108$.
9. $4c = 200$.
10. $50c = 100$.
11. $150c = 450$.
12. $12c = 144$.
13. $7c = 49$.
14. $16c = 48$.
15. $7c = 35$.
16. $11c = 121$.
17. $13c = 26$.
18. $13c = 39$.
19. $17c = 51$.
20. $9c = 81$.

Exercise 24

Find the value of y in each of the following equations.

1. $\dfrac{y}{2} = 4$
2. $\dfrac{y}{6} = 1$
3. $\dfrac{y}{8} = 3$
4. $\dfrac{y}{10} = 1$
5. $\dfrac{y}{8} = 3$
6. $\dfrac{y}{3} = 17$
7. $\dfrac{y}{5} = 10$
8. $\dfrac{y}{2} = 14$
9. $\dfrac{y}{3} = 1$
10. $\dfrac{y}{6} = 4$
11. $\dfrac{y}{7} = 3$
12. $\dfrac{y}{8} = 5$
13. $\dfrac{y}{7} = 8$
14. $\dfrac{y}{12} = 11$
15. $\dfrac{y}{12} = 12$
16. $\dfrac{y}{10} = 9$
17. $\dfrac{y}{9} = 8$
18. $\dfrac{y}{11} = 4$
19. $\dfrac{y}{6} = 11$
20. $\dfrac{y}{2} = 57$

Exercise 25

Find the value of the letter in each of the following equations.

1. $2a + 1 = 3$.
2. $5x + 3 = 13$.
3. $2a + 1 = 19$.
4. $11c - 3 = 30$.
5. $2t - 5 = 7$.
6. $3c + 5 = 17$.
7. $2d + 3 = 7$.
8. $3d + 1 = 19$.
9. $2d - 3 = 5$.
10. $7a + 1 = 15$.
11. $3p + 5 = 35$.
12. $2p + 11 = 25$.
13. $5a - 7 = 48$.
14. $2m + 3 = 101$.
15. $13x + 1 = 40$.
16. $13x - 1 = 38$.
17. $2d + 101 = 103$.
18. $14t + 1 = 29$.
19. $12a + 11 = 23$.
20. $12a - 11 = 1$.

Exercise 26

If $a = 2$ and $d = 4$ work number answers for:

1. $2a + d$.
2. $a + 2d$.
3. $2a - d$.
4. $6a + d$.
5. $6a - d$.
6. $3a - d$.
7. $\dfrac{a + d}{3}$
8. $\dfrac{2a + d}{8}$
9. $\dfrac{a + 3d}{7}$
10. $12a + \dfrac{d}{4}$.
11. $a + \dfrac{2d}{8}$.
12. $2(3a - d)$
13. $14a + d$.
14. $\dfrac{1}{4}(8a + d)$
15. $a + 14d$.
16. $12a - 2d$.
17. $\dfrac{18a + 3d}{16}$
18. $\dfrac{1}{2}a + \dfrac{1}{4}d$.
19. $5a + 11d$.
20. $\dfrac{8a + d}{20}$

Exercise 27

If $x = 1$, $y = 2$, $a = 3$ work number answers for:

1. $x + 2y$.
2. $y + 2a$.
3. $a + y + x$.
4. $2x + 3y$.
5. $4a + 2x$.
6. $\dfrac{2a}{y}$
7. $3(x + y - a)$.
8. $x^2 + y^2$.
9. $x + y^2$.
10. $2x + ya$.
11. $y^2 + xa$.
12. a^3.
13. $(a + y)^2$
14. $(2x + y)^2$.
15. $(x + 3a)^2$.
16. $\dfrac{2a - y}{8}$
17. $\dfrac{a^2 + 1}{5}$
18. $\dfrac{y^3 + 1}{3}$
19. $ay + ax$.
20. $a(x + y)$.

Exercise 28

In each of the following questions find the number:

1. Three times a certain number equals 201.
2. A number with 14 added to it equals 17.
3. Half of a certain number is 102.
4. One quarter of half of a certain number is 2.
5. A certain number multiplied by 10 equals 1000.

6. One seventh of a certain number is half of fourteen.
7. Twice a certain number, plus 7, equals 13.
8. Four times a certain number, less 3, equals 41.
9. Half of a number, halved again, gives 4.
10. Twelve times a certain number with 14 added equals 50.

Exercise 29

Write in simpler form:

1. $12ab \times 3$.
2. $22p \times 2$.
3. $\dfrac{14a}{2}$
4. $\dfrac{100x^2}{20}$
5. $\dfrac{18mnp}{9n}$
6. $2a \times 3m \times a$.
7. $12p^2q \times 3q$.
8. $\dfrac{20ab}{5a}$
9. $\dfrac{24mn}{12n}$
10. $\dfrac{4c^2d}{8d}$
11. $\dfrac{2at^2}{4}$
12. $2a \times 4b \times 6b$.
13. $12p \times 2p$.
14. $\dfrac{ab \times 4b}{8}$
15. $\dfrac{4cdm}{12c}$
16. $\dfrac{21a}{3}$
17. $\dfrac{2pq \times 4}{8p}$
18. $\dfrac{10c^2}{5}$
19. $2t \times 3t \times 6t$.
20. $\dfrac{4t^2}{8}$

Exercise 30

Say whether the following statements are true or false.

1. $5t + 1 = 16$ if $t = 3$.
2. $2m = 15$ if $m = 7\tfrac{1}{2}$
3. $3x - 1 = 5$ if $x = 2$.
4. $a + t = 14$ if $a = 11$ and $t = 3$.
5. $2a + 7 = 9$ if $a = 1$.
6. $3x - 4 = 5$ if $x = 3$.
7. $2t = 19$ if $t = 9$.
8. $3t + 5 = 11$ if $t = 2$.
9. $a + 2b = 14$ if $a = 2$ and $b = 6$.
10. $3t = 14$ if $t = 5$.
11. $2m + 7 = 9$ if $m = 2$.
12. $3a + c = d$ if $a = 3, b = 1, d = 10$.
13. $11p = 132$ if $p = 12$.
14. $17a = 35$ if $a = 2$.
15. $x + 11 = 140$ if $x = 129$.
16. $2x + 19 = 21$ if $x = 1$.
17. $a + t = 17$ if $a = 3$ and $t = 15$.
18. $14a = 29$ if $a = 1\tfrac{1}{2}$
19. $2p + 13 = 23$ if $p = 5$.
20. $100a = 1000$ if $a = 10$.

Arithmetic

Exercise 31

1. Find x if $\dfrac{x}{4} = 16$.
2. Find a if $2a - 1 = 43$.
3. Find c if $c - 9 = 4$.
4. Find t if $\dfrac{3t}{2} = 6$.
5. Find c if $\dfrac{c}{5} = 10$.
6. Find x if $2x - 13 = 4$.
7. Find x if $3x - 1 = 74$.
8. Find t if $\dfrac{3t}{4} = 15$.
9. Find g if $g - 1 = 3$.
10. Find x if $2x - 19 = 1$.
11. Find x if $\dfrac{5}{x} = 1$.
12. Find c if $\dfrac{10}{c} = 2$.
13. Find c if $\dfrac{11}{c} = 11$.
14. Find a if $7a - 3 = 4$.
15. Find x if $x + 100 = 102$.
16. Find t if $\dfrac{t}{17} = 2$.
17. Find a if $a + a + a = 6$.
18. Find x if $2x + 3x = 10$.
19. Find x if $7x + 3x = 40$.
20. Find a if $5a - 3a = 106$.

Exercise 32

1. x apples cost t pence. Write down the cost of 1 apple.
2. Four pencils cost £0·ta. Write down the cost of 1 pencil, in pence.
3. x books cost 50p. Write down the cost, in pence, of t books of the same kind.
4. There are p girls and q boys in a class. How many pupils are in the class?
5. There are m passengers on a bus. There is a driver and a conductor, also. How many people are on the bus when 1 person gets off?
6. y boys have each x pence to spend. How many pounds have they altogether?
7. x boys and y girls have £t each. How many pence is this, altogether?
8. A boy weighs x kg and a parcel weighing t kg is put into the bag. Write down the combined weight in grammes.
9. How many lengths of x cm each can be cut from a length of t metres?
10. 12 lengths each of t cm are cut from y metres of string. What length, in centimetres, remains?
11. How many bags each holding x kg can be filled from a sack containing p tonnes?
12. Put $2x$, $3y$, $5a$, $2d$ and $3e$ into ascending order if $x = 3$, $y = 2$, $a = 4$, $d = 5$ and $e = 1$.
13. Three chairs cost £a altogether. Write down the cost, in £s, of t chairs.
14. Find the total cost, in pence, of 3 pencils costing c pence each and x pens costing 25 pence each.

15. Express t cm in metres.
16. Express ac kg in tonnes.
17. Express k metres in mm.
18. A man walks for x hours at y km/h and then for t hours at p km/h. Write down the total distance walked, in kilometres.
19. A man works for x hours a day for 5 days and earns £t altogether. Write down his wage, per hour, in pence.
20. Express x mm^2 in cm^2.

Exercise 33

1. Express $2x$ km in metres.
2. A journey of t km is completed in 5 hours. At what rate was the distance covered?
3. p girls and q boys were in a room. 1 girl went out and 2 boys came in. How many people were then in the room?
4. A jar holds m kg of sweets. How often can d grammes of sweets be weighed out before the jar is empty?
5. Write down the value of $4p^2$ if $p = 1$.
6. Express $2p$ litres in cm^3.
7. A cuboid is x cm long, y cm wide and t cm high. Express its volume in m^3.
8. Add t metres and $2z$ km, giving your answer in metres.
9. x men, y women and p boys are in a hall. 7 girls come in. How many people are now in the room?
10. A piece of string measures c cm. How many centimetres of string will remain if three pieces each measuring t cm are cut off?
11. A man walks for m hours at p km/hour. How far has he travelled?
12. What is the next even number after $2t$?
13. What is the next even number after p if p is an odd number?
14. If p is divisible by 7 what is the next highest number, after p, which is also divisible by 7?
15. What number when multiplied by x gives t?
16. What number when divided by c gives d?
17. What number when added to 4 gives t?
18. What number when divided by 1000 gives q?
19. If ad is divisible by 28 and if $a = 4$ what is the smallest number that d can be?
20. If x kg is added to 1248 mg write down the total weight in grammes.

Exercise 34

Write in simpler form:

1. $\dfrac{2m \times 4}{8}$
2. $\dfrac{5p}{10}$
3. $\dfrac{3pq^2}{6q}$
4. $5ab \times 12$
5. $\dfrac{16a^2}{4}$
6. $\dfrac{17p}{34}$
7. $\dfrac{100c^2 d}{10c}$
8. $2pq \times 4$
9. $5at \times 2t$
10. $100p \times \tfrac{1}{2}p$
11. $\tfrac{1}{4} q \times 4p$
12. $10m \times \tfrac{1}{10} m$
13. $5p^2 \times 2p$
14. $\dfrac{12t^2}{3}$
15. $\dfrac{7eg^2}{14g}$
16. $2m^2 \times 4 \times 6m$
17. $\dfrac{3p \times 2q}{6q \times 12p}$
18. $\tfrac{1}{2} a \times \tfrac{1}{4} a$
19. $\dfrac{20d^2 \times 2m}{40m}$
20. $\dfrac{40x^2 y}{20x}$

Exercise 35

In each of the questions below find the value of the symbol.

1. $2m + 3 = m + 4$.
2. $5x - 2 = 4x + 6$.
3. $7x - 3 = 2x + 7$.
4. $4a - 2 = 3a + 7$.
5. $2x - 1 = x + 3$.
6. $4p - 3 = 2p + 7$.
7. $2m + 5 = m + 7$.
8. $3q - 1 = 2q + 4$.
9. $5q + 9 = 3q + 11$.
10. $5x + 17 = 3x + 21$.
11. $8x + 1 = 7x + 14$.
12. $6a - 19 = 5a + 1$.
13. $2x + 7 = x + 19$.
14. $4x + 13 = 3x + 14$.
15. $7p + 11 = 5p + 13$.
16. $2p - 1 = p + 3$.
17. $\dfrac{x}{2} + 1 = 3$.
18. $x + 18 = 19$.
19. $3x + 6 = 2x + 7$.
20. $2x + 7 = x + 15$.

SECTION 3

SOME PRACTICE WITH SIMPLE GEOMETRY PROBLEMS

Draw as carefullly as you can lines of lengths:

1.	4 cm	2.	8 cm	3.	6·5 cm	4.	9·5 cm	5.	1·5 cm
6.	35 mm	7.	105 mm	8.	59 mm	9.	72 mm	10.	48 mm
11.	14 mm	12.	6·3 cm	13.	9·1 cm	14.	8·8 cm	15.	4·6 cm

Measure the following lines giving your answer in mm.

16. A————————————————B

17. C——————————————————D

18. E————————————————F

19. G——————H

20. K————————L

21. M————N

22. P——————————————————Q

23. R——————————S

24. T——————————————————————U

25. X————————————————————Y

26. P————L

27. S————T

28. R——————W

29. N————————P

30. D————————E

A•————————•————————•————————•————E
       B   C   D

In the diagram AE is a straight line with B, C and D lying between A and E.

Give single line segments for the following:

31. AB + AC
32. AD + DE
33. AC + CD
34. AC + CE
35. CD + BC
36. BE + AB
37. AD − AB
38. BE − BC
39. BD − BC
40. AE − AB
41. BC + CD + AB
42. BE − DE
43. DE + BD − BC
44. DC + AC − AB
45. BC + AB + CD − AC

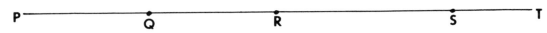

As before Q, R and S are points on a straight line PT.

If PT = 12 cm, PQ = 3·5 cm, QR = 2·5 cm, RT = 6 cm and ST = 1·5 cm.

Calculate the lengths of:

46. PR
47. PS
48. RS
49. QS
50. QT
51. PT − RT
52. PS − RS
53. QT − QR
54. QT − ST
55. PR + RS
56. QT − QR
57. TR + RQ
58. TQ − ST
59. PT − QP
60. TR − RS

61. Draw a diagram to illustrate the following. KM is a straight line 84 m long. J, L and N are points in the same straight line. J lies 22 m on the opposite side of K from M. N lies 14 m to the right of M and L lies between K and M 37 m from K.

What lengths are the following line segments?
62. JN
63. LM
64. LN
65. JL
66. JM
67. KN
68. KN − KL
69. JM − JK
70. JM + MN
71. LN − LM
72. JN − LN
73. JM − LM
74. KN − KL
75. KN − MN

Describe in terms of acute, obtuse, reflex, straight and right, angles of sizes:

76. 102°
77. 162°
78. 13°
79. 210°
80. 95°
81. 45°
82. 172°
83. 182°
84. 90°
85. 347°
86. 78°
87. 98°
88. 198°
89. 298°
90. 180°

In similar terms as above describe the angles between the hands of a clock showing the following times:

91. five past two
92. twelve fifteen
93. ten past five
94. quarter to seven
95. five to six
96. quarter to three
97. 2.35
98. 10.10
99. 11.25
100. 1.20

40 A Basic Mathematics Workbook

What size is the angle between the hands of a clock at the following times?

101. 1 o'clock	102. two o'clock	103. ten o'clock
104. nine o'clock	105. 5.15 p.m.	106. 11.15 p.m.
107. 4.40 p.m.	108. 1.55 a.m.	109. 6.10 p.m.
110. 8.10 a.m.	111. 7.05 p.m.	112. 10.15 p.m.
113. 3.05 a.m.	114. 9.10 p.m.	115. 2.20 a.m.

116. Line segments PQ and RS intersect at T such that ∠PTR = 42°. What size is (a) ∠STQ, (b) ∠PTS, (c) ∠RTQ?

117. Repeat above question for ∠PTS = 127°. What size is (a) ∠RTQ, (b) ∠STQ, (c) ∠PTR?

118.

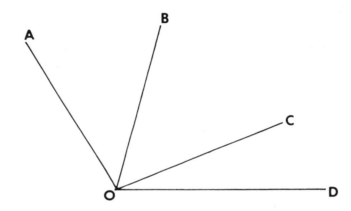

Name two angles adjacent to (a) ∠COD, (b) ∠BOC.

119. Give one angle to replace:
 (a) ∠BOA+∠BOC
 (b) ∠BOD+∠AOB
 (c) ∠AOD−∠BOD
 (d) ∠BOD−∠COD
 (e) ∠AOC+∠COD−∠AOB

120. In the diagram for question 118 ∠AOD = 110°, ∠AOB = 34° and ∠COD = 34°. Calculate the sizes of the following angles. (a) ∠BOC (b) ∠BOD (c) ∠AOC.

121. Still in the diagram for question 118 ∠AOB = 42°, ∠BOC = 34°, ∠COD = 39°. Calculate the size of the following angles. (a) ∠AOD (b) ∠AOC (c) ∠BOD.

Through what angle does the hand of a clock turn in:

122. 10 minutes	123. 20 minutes	124. 25 minutes
125. 40 minutes	126. 1 hour	127. 35 minutes
128. 45 minutes	129. 1½ hours	130. 2 hours
131. 8 minutes	132. 12 minutes	133. 41 minutes
134. 58 minutes	135. 65 minutes	

Calculate the supplements of the following angles:

136. 40°	137. 160°	138. 92°	139. 18°
140. 159°	141. 37°	142. 142°	143. 101°
144. 19°	145. 78°	146. 102°	147. 178°
148. 89°	149. 12°	150. 98°	

Calculate the complements of the following angles:

151. 45°	152. 75°	153. 81°	154. 11°	155. 24°
156. 38°	157. 42°	158. 63°	159. 29°	160. 69°
161. 15°	162. 4°	163. 37°	164. 52°	165. 58°

166. ∠AOB, ∠BOC, ∠COD and ∠DOA are four angles round AZ point. Calculate the fourth angle ∠AOD if:

(a) ∠AOB = 36° ∠BOC = 112° ∠COD = 84°
(b) ∠AOB = 97° ∠BOC = 128° ∠COD = 117°
(c) ∠AOB = 67° ∠BOC = 78°   ∠COD = 105°
(d) ∠AOB = 142° ∠BOC = 53°   ∠COD = 92°
(e) ∠AOB = 49° ∠BOC = 131° ∠COD = 56°
(f) ∠AOB = 104° ∠BOC = 23°   ∠COD = 167°
(g) ∠AOB = 73° ∠BOC = 122° ∠COD = 66°
(h) ∠AOB = 147° ∠BOC = 99°   ∠COD = 35°
(i) ∠AOB = 79° ∠BOC = 11°   ∠COD = 58°
(j) ∠AOB = 165° ∠BOC = 92°   ∠COD = 73°

167. Find the angle between each spoke of a wheel if they are equally spaced around the centre and there are:

(a) 12 spokes   (b) 18 spokes   (c) 20 spokes   (d) 36 spokes
(e) 40 spokes   (f) 45 spokes   (g) 60 spokes   (h) 80 spokes
(i) 72 spokes   (j) 120 spokes

168. Calculate the fifth angle of angles round a point where their sizes are:

(a) 42°, 74°, 79°, 83°, $x°$
(b) 116°, 28°, 67°, 124°, $y°$
(c) $p°$, 152°, 34°, 90°, 75°
(d) 132°, $q°$, 102°, 18°, 57°.

169. AB and CD are parallel lines cut by a transversal xy cutting AB in P and CD in Q. Draw a diagram and from it name:

(a) the angle vertically opposite ∠APX
(b) corresponding to ∠YQD
(c) alternate to ∠BPQ
(d) co-interior to ∠CQP

170. As in the diagram for question 169 if ∠APX = 73°, calculate:
(a) ∠PQC, (b) ∠PQD, (c) ∠XPB, (d) ∠APQ, (e) ∠CQY.

171.

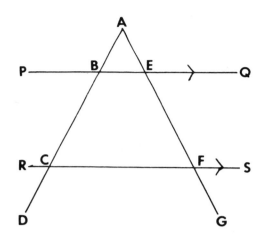

If ∠ABE = 52° and ∠GFS = 64° what size is:
(a) ∠BCF, (b) ∠RCB, (c) ∠GFC, (d) ∠AEB, (e) ∠BAE
(f) What can you say about ∠ABE?

172. Calculate the third angle of the following triangles if they have two angles of:
(a) 82°, 57°   (b) 46°, 76°   (c) 34°, 107°   (d) 51°, 64°
(e) 93°, 29°   (f) 73°, 78°   (g) 19°, 49°   (h) 114°, 25°
(i) 62°, 43°   (j) 42°, 87°

173. Calculate the third angle of the following right angled triangles:
(a) 18°   (b) 63°   (c) 23°   (d) 51°
(e) 32°   (f) 76°   (g) 83°   (h) 7°
(i) 45°   (j) 49°

174. Calculate the size of the equal angles in an isosceles triangle where the third angle is:
(a) 36°   (b) 24°   (c) 56°   (d) 100°
(e) 92°   (f) 124°   (g) 68°   (h) 84°
(i) 76°   (j) 112°

175. Calculate the size of the remaining angle of an isosceles triangle where the equal angles are:
(a) 42°   (b) 65°   (c) 53°   (d) 27°
(e) 74°   (f) 38°   (g) 82°   (h) 75°
(i) 48°   (j) 68°

176. Construct the following triangles from the information given:
(a) AB = 5 cm, AC = 6·5 cm, ∠BAC = 58°
(b) PQ = 7 cm, PR = 4·8 cm, ∠QPR = 44°
(c) DE = 10 cm, DF = 8·4 cm, ∠FDE = 72°
(d) XY = 9·5 cm, YZ = 7·4 cm, ∠XYZ = 116°

Simple Geometry Problems

177. Indicate whether a triangle could have the following angle sizes:
   (a) 34°, 76°, 70°   (b) 71°, 63°, 56°   (c) 81°, 79°, 20°
   (d) 107°, 44°, 29°   (e) 69°, 57°, 54°   (f) 29°, 111°, 40°

178. In the following right angled triangles state (1) the hypotenuse (2) the smallest side:
   (a) $\triangle DEF$, $\angle E = 90°$, $\angle F = 40°$
   (b) $\triangle KLM$, $\angle K = 90°$, $\angle L = 62°$
   (c) $\triangle RST$, $\angle T = 90°$, $\angle S = 54°$
   (d) $\triangle WXY$, $\angle W = 90°$, $\angle Y = 32°$
   (e) $\triangle GHJ$, $\angle H = 90°$, $\angle J = 47°$
   (f) $\triangle BCD$, $\angle D = 90°$, $\angle B = 29°$

179. ABCD is a rectangle whose diagonals intersect at E. Draw a diagram. State whether the following are true or false.
   (a) AB = BC   (b) AB = CD   (c) AC = BD
   (d) AE = EC   (e) AE = BE   (f) DB = BE
   (g) $\angle BAC = \angle DAC$ (h) $\angle DAE = \angle BEC$
   (i) $\angle ABD + \angle DBC = 90°$ (j) $\angle AED + \angle BEC = 180°$
   (k) $\angle ABD = \angle BDC$ (l) $\angle ADB = \angle ABD = 90°$
   (m) $\angle DAE = \angle ADE$ (n) $\angle DEC + \angle BEC = 180°$
   (o) $\angle EAB = \angle EBA$

180. Is every rectangle a square?

181. From the properties of a square which of the following are true/false?
   (a) it has only 4 axes of symmetry.
   (b) the diagonals intersect at right angles.
   (c) the diagonals bisect the angles through which they pass.
   (d) fits in its outline four ways.
   (e) the diagonals divide it into four congruent equilateral triangles.

182. In a rectangle PQRS the diagonals meet at T if PQ = 12 cm and PS = 5 cm and PR = 13 cm. Calculate the lengths of:
   (a) QR, (b) SR, (c) SQ, (d) PT, (e) ST

183. If in the rectangle of question 182 $\angle PQS = 35°$, calculate:
   (a) $\angle QPR$ (b) $\angle PTQ$ (c) $\angle PTS$ (d) $\angle SQR$ (e) $\angle QTR$ (f) $\angle QSR$

184. Rectangle KLMN has diagonals intersecting at O. If $\angle KOL$ is 126° calculate:
   (a) $\angle NOM$ (b) $\angle LNM$ (c) $\angle KMN$ (d) $\angle KNL$ (e) $\angle MLN$ (f) $\angle LOM$

185. Construct the following triangles from the given data.
   (a) $\triangle ABC$ where AB = 7 cm, $\angle A = 46°$, $\angle B = 57°$
   (b) $\triangle MNP$ where NP = 8·5 cm, $\angle N = 84°$, $\angle P = 38°$
   (c) $\triangle POQ$ where PQ = 6·2 cm, $\angle P = 103°$, $\angle Q = 42°$
   (d) $\triangle DEF$ where DE = 9 cm, $\angle D = 64°$, $\angle E = 69°$
   (e) $\triangle SRT$ where RT = 11·5 cm, $\angle R = 72°$, $\angle T = 28°$

186.

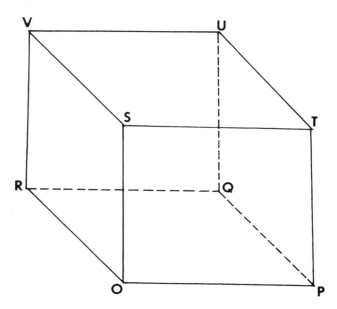

OPQR, STUV is a cuboid as shown
(a) Name four vertical parallel lines
(b) Four horizontal parallel lines

(c) a rectangle congruent to (1) OSVR (2) RQUV
(d) a line equal to (1) SU (2) RS (3) OT
(e) a rectangle congruent to (1) OSUQ (2) VSPQ

If OP = 8 cm, PQ = 6 cm and TP = 5 cm
(f) what is the perimeter of rectangle (1) OPQR (2) OPTS (3) OSVR?
(g) what is the area of the rectangles in question (f)?
(h) what is the total length of the edges?
(i) what is the total surface area?
(j) what is the volume of OPQR, STUV?
(k) what three other lines are equal to QS?

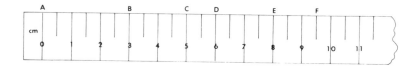

What are the lengths of the following?

187. AB	188. AC	189. AD	190. AE
191. AF	192. BC	193. BD	194. BE
195. BF	196. CD	197. CE	198. CF
199. DE	200. DF	201. EF	202. AC + DE
203. BC + EF	204. BD + CE	205. AD + DF	206. CF + AB
207. AD − BC	208. BF − DE	209. AC − CD	210. CF − BC
211. AF − BE			

Simple Geometry Problems   45

212.

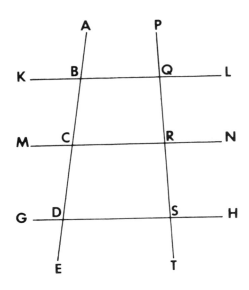

KLMN and GH are parallel lines.
∠ABK = 115° and ∠HST = 84°. What sizes are:
(a) ∠CBQ   (b) ∠MCB   (c) ∠MCD
(d) ∠BCR   (e) ∠RQL   (f) ∠CRQ
(g) ∠RSD   (h) ∠QRN?

213. RSTV is a square whose diagonals intersect at O. RS = 10 cm and RT = 14 cm. Calculate:
    (a) the perimeter of RSTV
    (b) the area of RSTV
    (c) the length of OS
    (d) length of SV
    (e) the size of ∠ORS
    (f) the size of ∠VOR
    (g) the area of △VRS
    (h) area of △VOR
    (i) is △TOS congruent to △VOR?
    (j) is △ROS congruent to △SOT?

214. Is every cube a cuboid?

215.

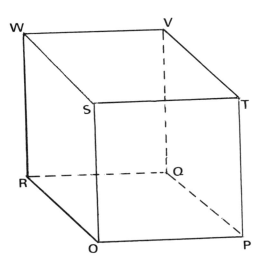

OPQR, STVW is a cube with OP = 6 cm. Calculate the following:

(a) perimeter of OPST
(b) perimeter of STVW
(c) the total length of the edges
(d) area of OSWR
(e) total surface area
(f) volume of the cube
(g) size of ∠TOP
(h) ∠SVT
(i) name three sets of 4 parallel lines
(j) draw 2 different nets of the cube.

46 A Basic Mathematics Workbook

216. What is the size of the angle between the following compass points? (Take the smaller angle.)
(a) N. and S.E.
(b) N. and N.W.
(c) N.E. and S.W.
(d) N.E. and S.E.
(e) E. and S.W.
(f) E. and N.W.
(g) S. and S.W.
(h) S.W. and N.W.
(i) W. and N.E.
(j) N.W. and S.W.

217. Give the following 3 figure bearings in terms of compass bearings.
(a) 045° (b) 180° (c) 225° (d) 315° (e) 135°

218. Give the following compass bearings in terms of 3 figure bearings (e.g. N.40°W becomes 320°).
(a) N.48°E. (b) S.20°E. (c) S.20°W. (d) N.70°W. (e) N.65°E.
(f) S.35°W. (g) N.55°E. (h) S.63°E (i) S.82°W. (j) N.74°W.

219. Rewrite the following 3 figure bearings as compass directions like those in the previous question (e.g. 030° becomes N.30°E.).
(a) 120° (b) 210° (c) 050° (d) 300° (e) 240°
(f) 075° (g) 165° (h) 280° (i) 325° (j) 015°

220. The bearing of point A from point B is 040°. What is the bearing of point B from point A?

221. The bearing of point P from point Q is 200°. What is the bearing of point Q from point P?

222. What is the complement of 46°? What is the supplement of your answer?

223. What is the supplement of 112°? What is the complement of your answer?

224. In figure 1, PQ is parallel to RS, $\angle PRT = 130°$ and $\angle QPR = 75°$. Calculate:
(a) $\angle PRQ$ (b) $\angle PQR$ (c) $\angle QRS$ (d) $\angle TRS$

225. In figure 2, DE = DF, DH = HG and $\angle DHG = 46°$. Calculate:
(a) $\angle HDG$ (b) $\angle DFE$.

226. In figure 3, AB is parallel to CD, $\angle ABC = 72°$ and $\angle ACB = 42°$. Calculate:
(a) $\angle BAC$ (b) $\angle DCE$ (c) $\angle BCD$.

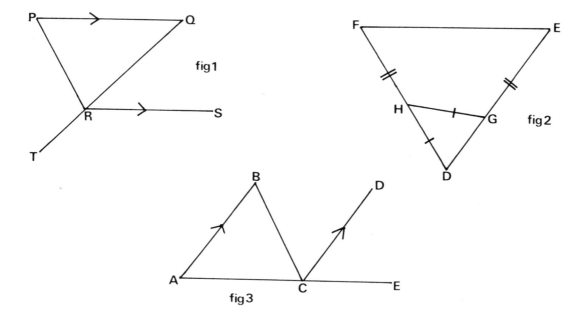

Simple Geometry Problems 47

227. In figure 4, WV is parallel to YX, ∠VWZ = 83° and ∠WVZ = 74°. Calculate:
 (a) ∠XYZ (b) ∠YXZ (c) ∠XZY

228. In figure 5, PQ is parallel to RS.
 (a) what angle is vertically opposite ∠GKS?
 (b) what angle is the supplement of ∠PIJ?
 (c) what angle corresponds with ∠GHK?
 (d) what angle is alternate to ∠ILK?
 (e) what angle is co-interior with ∠HIL?

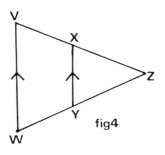

fig 4

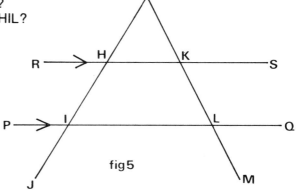

fig 5

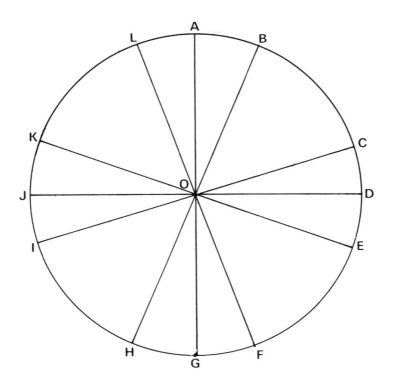

Which angle is vertically opposite the given angle?

48 A Basic Mathematics Workbook

230. ∠AOB   231. ∠AOC   232. ∠AOD   233. ∠KOL   234. ∠KOJ
235. ∠KOH   236. ∠COD   237. ∠DOF   238. ∠FOH   239. ∠FOL
240. ∠GOJ

∠AOB = ∠AOL = 18°; ∠BOC = ∠KOL = 48°; ∠COD = ∠KOJ = 24°.

Using above information (or similar).
Find the sizes of the following angles.

241. ∠GOF   242. ∠HOF   243. ∠AOC   244. ∠BOD   245. ∠COE
246. ∠BOE   247. ∠COL   248. ∠COF   249. ∠LOD   250. ∠AOD
251. ∠BOK   252. ∠KOC   253. ∠KOD   254. ∠JOC   255. ∠GOK
256. ∠GOL   257. ∠GOC   258. ∠KOF   259. ∠AOF   260. ∠AOI
261. ∠AOE   262. ∠AOJ   263. ∠AOF   269. ∠AOH   265. ∠JOE
266. ∠JOC   267. ∠FOJ   268. ∠HOE   269. ∠GOB

Which of the following groups of three angles could form the angles of a triangle?

270. 56°, 29°, 95°   271. 112°, 42°, 25°   272. 78°, 59°, 33°
273. 45°, 68°, 67°   274. 43°, 85°, 52°   275. 37°, 41°, 92°
276. 102°, 57°, 21°   277. 96°, 23°, 61°   278. 54°, 19°, 117°
279. 18°, 121°, 41°

280. PQR is a straight line. Form an equation in x and y. Use your equation to find values for x when y has the value:

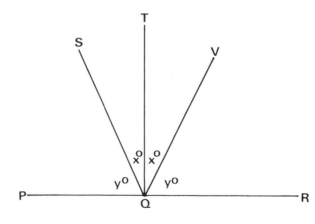

281. 60°   282. 50°   283. 45°   284. 70°   285. 52°
286. 57°   287. 63°   288. 66°   289. 71°   290. 78°
291. 48°   292. 54°   293. 69°   294. 74°   295. 81°

Use your equation to find values for y when x has the value:
296. 10°   297. 30°   298. 40°   299. 15°   300. 25°

Simple Geometry Problems 49

301.	35°	**302.**	12°	**303.**	18°	**304.**	23°	**305.**	29°
306.	32°	**307.**	36°	**308.**	7°	**309.**	17°	**310.**	39°

311. If the value of x is increased by 10° what happens to the value of y?

312. If the value of y is decreased by 20° what happens to the value of x?

313. If $\angle ABC = 90°$ form an equation in p and q.

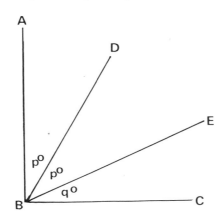

Use your equation to find the value of p if q has the value:

314.	30°	**315.**	20°	**316.**	40°	**317.**	14°	**318.**	24°
319.	36°	**320.**	16°	**321.**	32°	**322.**	28°	**323.**	52°
324.	62°	**325.**	76°	**326.**	58°	**327.**	44°	**328.**	22°

Use your equation to find the value of q if p has the value:

329.	10°	**330.**	15°	**331.**	20°	**332.**	40°	**333.**	35°
334.	8°	**335.**	12°	**336.**	17°	**337.**	21°	**338.**	28°
339.	31°	**340.**	41°	**341.**	37°	**342.**	18°	**343.**	27°

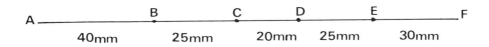

The distances between the points are shown. Calculate the lengths of the following:

344.	AC	**345.**	AD	**346.**	AE	**347.**	AF	**348.**	BD
349.	BE	**350.**	BF	**351.**	CE	**352.**	CF	**353.**	DF
354.	AB + CD	**355.**	AB + DE	**356.**	AB + EF	**357.**	BC + DE		
358.	BC + EF	**359.**	AB + CD + EF	**360.**	CD + EF	**361.**	AF − EF		
362.	BF − EF	**363.**	AD − CD	**364.**	AE − DE	**365.**	AE − CE		
366.	BE − DE	**367.**	BF − BC	**368.**	CF − CD	**369.**	AF − BE		

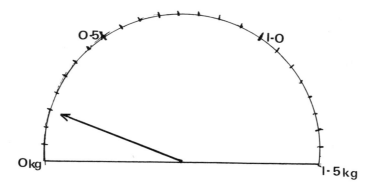

370. A set of kitchen scales are calibrated as shown, e.g. the pointer moves through a semi circle in measuring a weight of 1·5 kg. Through how many degrees will the pointer turn in weighing:

(a) 1 kg? (b) 0·5 kg? (c) 0·8 kg? (d) 1·2 kg? (e) 0·4 kg?
(f) 1·4 kg? (g) 100 g? (h) 300 g? (i) 600 g? (j) 700 g?
(k) 900 g? (l) 1300 g? (m) 250 g? (n) 750 g? (o) 150 g?
(p) 450 g? (q) 1250 g? (r) 1450 g? (s) 850 g? (t) 950 g?

What will the scale measure if the pointer turns through (give the answers in gm):
(a) 120°? (b) 30°? (c) 150°? (d) 90°? (e) 72°?
(f) 36°? (g) 144°? (h) 108°? (i) 96°? (j) 12°?
(k) 6°? (l) 66°? (m) 174°? (n) 32°? (o) 48°?
(p) 126°? (q) 156°? (r) 54°? (s) 78°? (t) 162°?

371. A ship sails from A to B on a course of 042°. What course must be set to return from B to A?

372. An aeroplane flies due North for 100 km. It then changes direction and flies due East for 100 km. What direction must it take to fly back to its original position?

373.

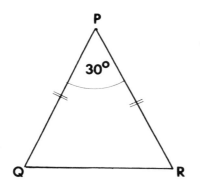

Find ∠PQR.
What can you say about △PQR?

374.

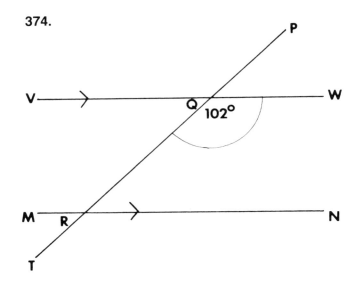

Find all the angles in the figure.

375.

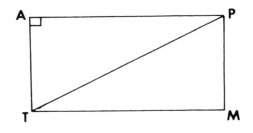

APMT is a rectangle.
AP = 6 cm
AT = 2·5 cm

Give lengths of: (1) PM, (2) TM
What name is given to line TP?
What can you say about the lines TP and AM?

376.

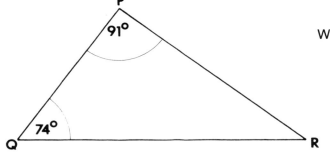

What size is ∠PRQ?

52 A Basic Mathematics Workbook

377.

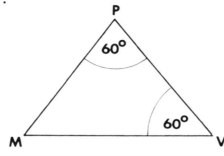

What size is ∠PMV?
What can you say about triangle PMV?

378.

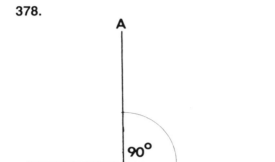

If CBD is a straight line, what size is ∠ABC?

379.

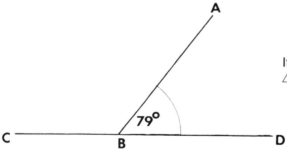

If CBD is a straight line, what size is ∠ABC?

380.

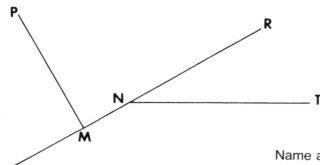

Name as many line segments as you can from the diagram.

381.

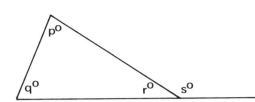

Calculate r and s if p and q have the following values:
(a) p = 42, q = 56   (b) p = 48, q = 62
(c) p = 54, q = 56   (d) p = 64, q = 70
(e) p = 63, p = 72   (f) p = 29, q = 82
(g) p = 39, q = 67   (h) p = 43, q = 68
(i) p = 51, q = 73   (j) p = 61, q = 38

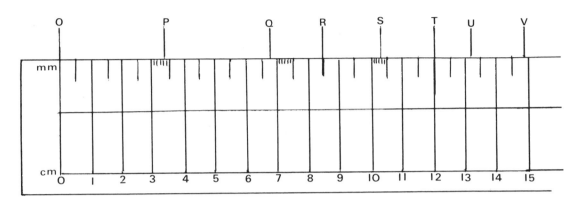

Give the lengths of the following line segments in cm.
383. OP	383. OQ	384. OR	385. OS	386. OT
387. OV	388. VU	389. VO	390. VT	391. VS
392. VR	393. VQ	394. VP	395. PQ	396. PR
397. PS	398. PT	399. PU	400. PV	401. QR
402. QS	403. QT	404. QU	405. QV	

Give the following lengths in mm.
406. RS	407. RT	408. RU	409. RV
410. ST	411. SU	412. SV	413. TU
414. TV	415. OP + VU	416. OQ + VT	417. OR + VS
418. QP + QS	419. SR + ST	420. TR + TV	421. RP + RU

422.

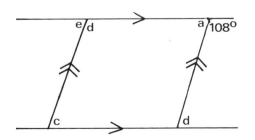

Calculate the sizes of the angles marked a, b, c, d, e.

423.

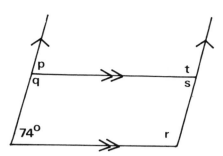

Calculate the sizes of the angles marked p, q, r, s, t.

424. In the previous question what kind of angles are:
(a) *p* and *s* (b) *r* and *t* (c) *q* and *s* ?

425.

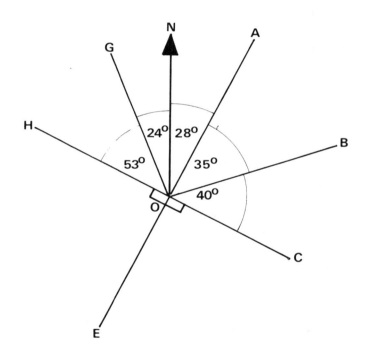

From the diagram give the three figure bearing of:
(a) A from 0 (b) H from 0
(c) B from O (d) G from O
(e) C from O (f) E from O

426. What is the compass direction of:
(a) B (b) E (c) D (d) H (e) G
(f) F (g) A (h) C from the point 0?

427.

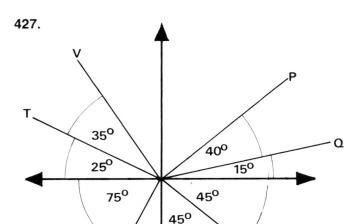

What is the compass direction of:

(a) P (b) Q (c) R
(d) S (e) T (f) V from O?

428. What are the three figure bearings of O from:
(a) P (b) Q (c) T?

429. What are the three figure bearings of:
(a) R (b) S (c) V from?

SECTION 4

GEOMETRY EXERCISES

Exercise 1

Draw lines of:

1. 13 mm	2. 2·9 inches	3. 5·2 cm
4. 28 mm	5. 4·1 inches	6. 7·4 cm
7. 37 mm	8. 3·2 inches	9. 6·8 cm
10. 49 mm	11. 1·7 inches	12. 8·2 cm
13. 58 mm	14. 3·8 inches	15. 11·6 cms
16. 91 mm	17. 4·6 inches	18. 8·8 cm
19. 42 mm	20. 3 cm	

Exercise 2

Say whether the following angles are acute, obtuse or reflex.

1. 32°	2. 104°	3. 341°	4. 17°
5. 92°	6. 44°	7. 186°	8. 132°
9. 191°	10. 312°	11. 4°	12. 104°
13. 181°	14. 93°	15. 87°	16. 100°
17. 359°	18. 310°	19. 11°	20. 187°

Exercise 3

Name as many line segments as you can see in the following figures.

1.

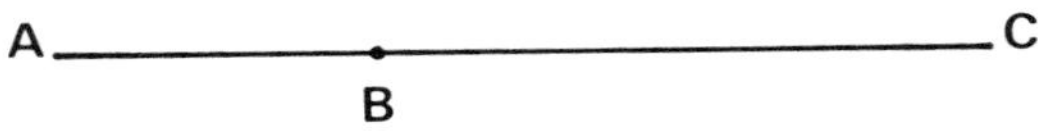

A————•————C
       B

2.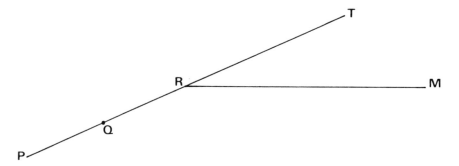

3.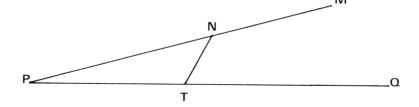

4.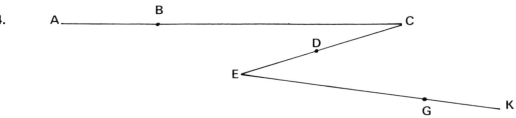

5.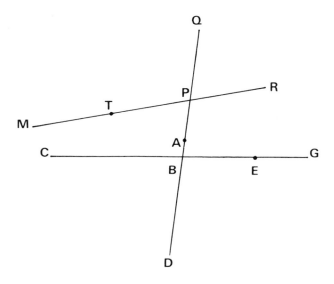

Exercise 4

In each of the examples 1 to 20 the sizes of 2 angles of a triangle are given. Work out the size of the third angle in each case.

1. 22°; 100°
2. 44°; 102°
3. 81°; 72°
4. 44°; 44°
5. 16°; 96°
6. 108°; 18°
7. 30°; 30°
8. 60°; 60°
9. 90°; 45°
10. 103°; 21°
11. 51°; 68°
12. 39°; 48°
13. 84°; 61°
14. 22°; 22°
15. 63°; 47°
16. 28°; 34°
17. 42°; 49°
18. 19°; 99°
19. 47°; 47°
20. 100°; 41°

Exercise 5

Draw a sketch to show each of the following directions.

1. 040°
2. N.W.
3. 031°
4. 121°
5. S.W.
6. N.N.W.
7. 226°
8. N.E.
9. N.N.E.
10. 032°
11. 141°
12. S.S.W.
13. 016°
14. 330°
15. 218°
16. S.E.
17. 82°
18. E.S.E.
19. 196°
20. 310°

Exercise 6

Find the perimeters of the rectangles whose lengths and breadths are given below.

1. $l = 2$ cm; $b = 1$ cm
2. $l = 5$ cm; $b = 2$ cm
3. $l = 14$ cm; $b = 2$ cm
4. $l = 10$ cm; $b = 8$ cm
5. $l = 12$ cm; $b = 8$ cm
6. $l = 7$ cm; $b = 5$ cm
7. $l = 100$ cm; $b = 90$ cm
8. $l = 90$ cm; $b = 70$ cm
9. $l = 2$ m; $b = 1\frac{1}{2}$ m
10. $l = 3$ m; $b = 1$ m
11. $l = 44$ cm; $b = 21$ cm
12. $l = 38$ cm; $b = 16$ cm
13. $l = 50$ cm; $b = 25$ cm
14. $l = 27$ cm; $b = 18$ cm
15. $l = 4$ m; $b = 3$ m
16. $l = 2·2$ m; $b = 1·6$ m
17. $l = 3·8$ cm; $b = 1·8$ cm
18. $l = 4·9$ cm; $b = 4·6$ cm
19. $l = 14·2$ cm; $b = 12$ cm
20. $l = 8·2$ cm; $b = 7·1$ cm

Exercise 7

Find the perimeters and areas of the squares whose side lengths are given below.

1. 2 cm
2. 3·8 cm
3. 6 mm
4. 1 m
5. 6 cm
6. 4·2 cm
7. 20 mm
8. 1·5 m
9. 4 cm
10. 3·1 cm
11. 10 mm
12. 1·2 m
13. 5·5 cm
14. 5·8 cm
15. 11 mm
16. 1·1 m
17. 3·8 cm
18. 2·9 cm
19. 18 mm
20. 2·2 m

Exercise 8

In the following diagrams, find the sizes of the unmarked angles.

1.

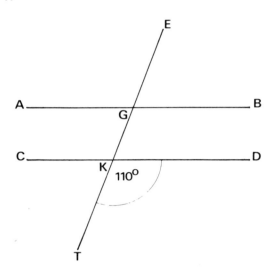

AB and CD are parallel lines.

2.

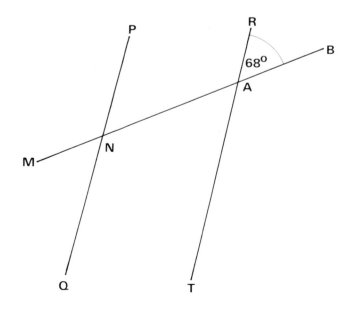

PQ and RT are parallel lines.

3.

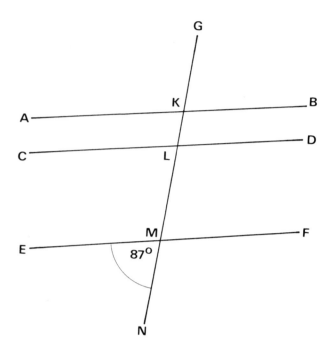

AB, CD and EF are parallel lines.

4.

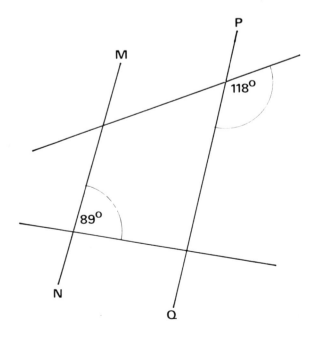

MN and PQ are parallel lines.

Exercise 9

In each of the questions below, draw △ABC, using the given data:
1. AB = 3 cm; BC = 5 cm; ∠ABC = 24°. Measure AC.
2. AC = 8 cm; AB = 5 cm; ∠BAC = 100°. Measure BC.
3. AC = 6 cm; BC = 12 cm; ∠ACB = 30°. Measure AB.
4. AB = 3·8 cm; BC = 6·3 cm; ∠ABC = 49°. Measure ∠BAC.
5. AC = 7·6 cm; AB = 4·8 cm; ∠BAC = 93°. Measure ∠BCA.
6. AC = 4·8 cm; BC = 11·2 cm; ∠ACB = 41°. Measure ∠ABC.
7. AB = 8 cm; BC = 9 cm; AC = 7·2 cm. Measure ∠ABC.
8. AB = 7·4 cm; BC = 10·8 cm; AC = 8·1 cm. Measure ∠ACB.
9. AB = 3·2 cm; BC = 3·2 cm; AC = 2·8 cm. What can you say about △ABC?
10. AB = 4·8 cm; AC = 5·6 cm; BC = 7 cm.

Exercise 10

Draw △PQR in each of the following questions, using the given data:
1. PQ = 8 cm; ∠QPR = 42°; ∠PQR = 56°. Measure PR.
2. QR = 10 cm; ∠PQR = 98°; ∠PRQ = 38°. Measure PR.
3. QR = 8 cm; ∠PRQ = 41°; ∠PQR = 62°. Measure PQ.
4. PQ is at right-angles to QR; PQ = 2·5 cm; QR = 4·8 cm. Measure PR.
5. PR is at right-angles to QR; QR = 10 cm; PR = 8 cm. Measure PQ.
6. QR = 8·8 cm; ∠PQR = 49°; ∠PRQ = 78°. Measure PR.
7. PQ = 8 cm; PR = 6 cm; ∠QPR = 52°. Measure QR.
8. QR = 12 cm; PR = 8 cm; ∠PRQ = 84°. Measure PQ.
9. PQ = 3·8 cm; QR = 4·9 cm; PR = 5·1 cm. Measure ∠PQR.
10. QR = 11 cm; PQ = 10·8 cm; ∠PRQ = 50°. Measure PR.

Exercise 11

Using your ruler and protractor, draw angles of:

1. 32°	2. 330°	3. 146°	4. 29°
5. 119°	6. 115°	7. 131°	8. 79°
9. 312°	10. 284°	11. 71°	12. 99°
13. 105°	14. 198°	15. 210°	16. 342°
17. 339°	18. 17°	19. 82°	20. 120°

Exercise 12

Copy the following figures. **Write down** the measurements asked for in each case.

1.

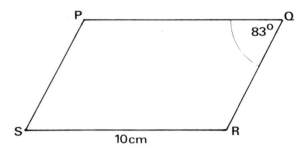

PQRS is a parallelogram.
Give the sizes of:

(a) PQ
(b) PS
(c) ∠PSR
(d) ∠SPQ
(e) ∠SRQ

2.

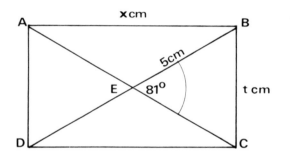

ABCD is a rectangle.
Give the sizes of:

(a) AE
(b) EC
(c) DE
(d) ∠AED
(e) AD
(f) DC
(g) ∠DEC

3.

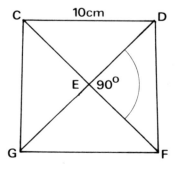

CDFG is a square.
Write down the sizes of:

(a) GE
(b) CE
(c) EF
(d) DF
(e) GF
(f) CG
(g) ∠GEF

4.

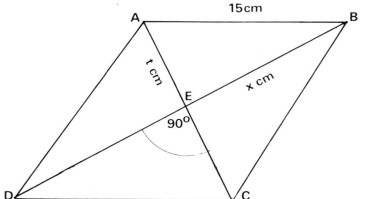

ABCD is a rhombus.
Write down the sizes of:

(a) BC
(b) DC
(c) AD
(d) ∠AEB
(e) DE
(f) EC

Exercise 13

In the following questions find the sizes, where required, by studying the diagrams.

1.

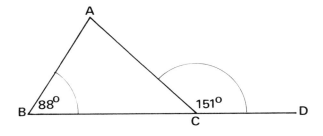

Find (a) ∠BAC
Find (b) ∠ACB

2.

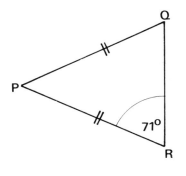

Find (a) ∠PQR
Find (b) ∠QPR

3.

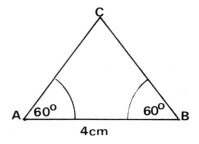

Find (a) ∠ACB
Find (b) BC

4.

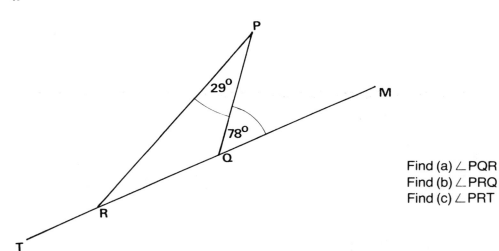

Find (a) ∠PQR
Find (b) ∠PRQ
Find (c) ∠PRT

5.

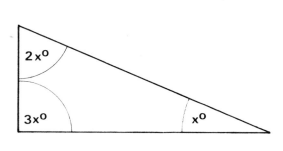

Find (a) x, in degrees.
Find (b) ∠ABC
Find (c) ∠ACB
Find (d) ∠BAC

Exercise 14

In each case, give the angle which is the **complement** of:

1. 31°
2. 26°
3. 74°
4. 82°
5. 1°
6. 17°
7. $14\frac{1}{2}°$
8. 72°
9. $10\frac{1}{2}°$
10. 89°
11. 60°
12. 30°
13. 45°
14. $47\frac{1}{2}°$
15. 49°
16. 12°
17. $9\frac{1}{2}°$
18. 83°
19. $77\frac{1}{2}°$
20. 65°

Exercise 15

In each case, give the angle which is the **supplement** of:

1. 3°
2. 8°
3. 14°
4. 171°
5. 90°
6. 89°
7. 114°
8. 178°
9. 91°
10. 111°
11. 164°
12. 5°
13. $7\frac{1}{2}°$
14. $112\frac{1}{2}°$
15. 29°
16. 129°
17. 138°
18. $2\frac{1}{2}°$
19. 179°
20. $100\frac{1}{2}°$

Exercise 16

Construct the following figures, using the given data.

1. Draw rectangle ABCD with AB = 5 cm; BC = 3 cm.
2. Draw square PQRS with RS = 8 cm.
3. Draw square ABCD with AC = 10 cm.
4. Draw rhombus PQRS with PQ = 6 cm; ∠PQR = 80°.
5. Draw rectangle ABCD with DC = 9 cm; CB = 4 cm.
6. Draw parallelogram ABCD with AB = 6 cm; BC = 4 cm; ∠ABC = 63°.
7. Draw square PQRS with SQ = 12 cm.
8. Draw rhombus ABCD with DC = 6 cm; ∠DAB = 72°.
9. Draw square PQRS with PQ = 5 cm.
10. Draw parallelogram PQRS with QR = 8 cm; SR = 12 cm; ∠SPQ = 68°.

Exercise 17

1.

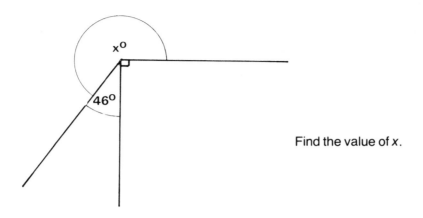

Find the value of x.

2.

State the sizes of:
(a) ∠ABE
(b) ∠EBC

3.

PQRS is a parallelogram.

Give the size of:
(a) QR
(b) SR
(c) ∠PQR

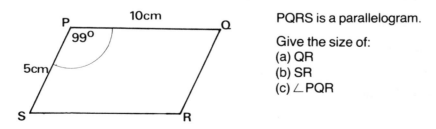

4. In question **3**, work out the area, in square centimetres, of parallelogram PQRS.

5.

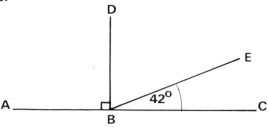

What size is ∠DBE?

6.

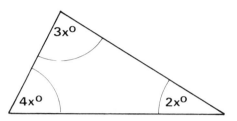

Find x and the size of each angle of the triangle.

7. Draw a triangle, TMP, with TM = 3·8 cm, MP = 6·9 cm and ∠TMP = 86°. Measure TP.

8. How many degrees would be turned through in three complete revolutions?

9.

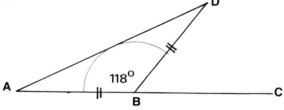

Show that
∠DBC equals the sum of
∠BDA and
∠DAB.

10.

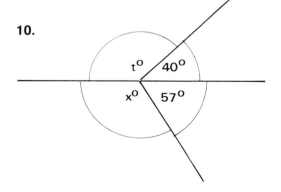

By how many degrees is t bigger than x?

Exercise 18

1. Sketch a cuboid of length 4 cm, breadth 3 cm and height 2 cm.
2. Sketch a cube of side length 4 cm.
3. Sketch a cube of side length 3 cm.
4. Sketch a cube of side length 2 cm.
5. Find in cubic centimetres the total volume of the solids referred to in questions **1, 2, 3** and **4**.
6. What is the total surface area of the cuboid in question **1**?
7. What is the total surface area of the cube referred to in question **2**?
8. What is the total surface area of the cube referred to in question **3**?
9. What is the total surface area of the cube referred to in question **4**?
10. How many cubes of side length $\frac{1}{2}$ cm would make a cube of the same volume as the cube in question **4**?

Exercise 19

1. Sketch an equilateral triangle of side length 4 cm.
2. What size is each angle of the triangle in question **1**?
3. Sketch an isosceles triangle with a base length of 5 cm and with its equal sides each 6 cm long.
4. Repeat the sketch of question **3** and then draw another isosceles triangle with its equal sides each 6 cm long on the opposite side of the base line from your first triangle.
5. What kind of figure is the combined figure you have obtained in question **4**?
6. Draw a rhombus of side length 6 cm with one of its angles equal to 40°.
7. Are the diagonals of the rhombus of question **6** equal in length?
8. Draw a square which will have the same area as a rectangle of length 24 cm and width 6 cm. State the side length of your square. Measure the diagonals of your square. Show that they are each almost 17 cm long.
9. A map is drawn to scale of $\frac{1}{100\,000}$. What length, in kilometres, is a road which is 0·32 cm long on the map?
10. What do you know about the diagonals of a rhombus at the point where they intersect? Give **two** facts.

Exercise 20

1.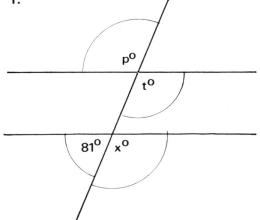

What can you say about x, t and p?

2.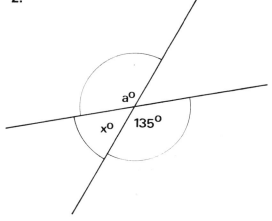

Find x and a.

3.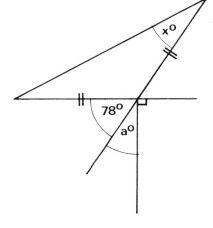

Find x and a.

4.

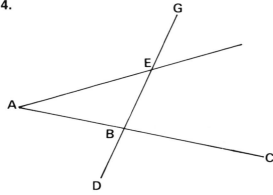

How many line segments can you name from the figure on the left?

5.

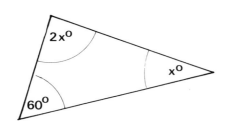

Find x and give the sizes of the angles of the triangle on the left.

6.

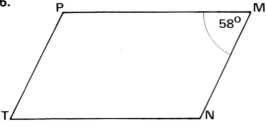

PMNT is a parallelogram.
(a) What length is PT?
(b) What length is TN?
(c) What length is MN?
(d) What size is ∠TNM?
(e) What size is ∠TPM?

7.

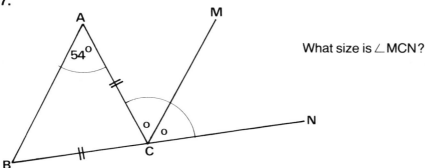

What size is ∠MCN?

8.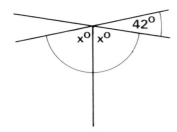

Find the value of x.

9. Construct △PQR, with PQ = 8 cm, ∠PQR = 44° and ∠QPR = 58°. Measure QR.

10. Construct △XYZ with XY = 4·6 cm, YZ = 7·8 cm and ∠XYZ = 51°. Measure XZ.

SECTION 5

500 MISCELLANEOUS QUESTIONS

Work answers to the following questions:

1. A man bought a car for £1100 and sold it for £950. How much did he lose?
2. If 29 out of 10 000 people questioned do not watch sport on TV, what percentage is this?
3. How much greater than $2\frac{1}{2}$ is $1\frac{3}{4} \times 2\frac{7}{8}$?
4. What interest has to be paid on £1000 borrowed for 8 months at 12% per annum simple interest?
5. Which is smaller; $\frac{3}{7}$ or $\frac{5}{14}$?
6. What is the average of 15, 19, 23, 24 and 34?
7. An article which cost £4 had its price increased by 10%. What was the new price?
8. How many metres are there in 28·4 km?
9. An article costing £100 cash is sold on Hire Purchase for a 20% deposit and 12 equal payments of £7·40. How much extra was charged for Hire Purchase?
10. How many litres of water can a cuboid hold if its internal measurments are 3 m, $2\frac{1}{2}$ m and $1\frac{3}{4}$ m?
11. If $a = 2$ and $x = 4$ what is the value of $2(3x + 5a)$?
12. Find x if $2x - 1 = 19$.
13. How many eggs can be bought for £1 if t eggs cost x pence altogether?
14. Give a simpler answer for $3a + 5a - 2a + 7a$.
15. Find t if $3t - 5 = 1$.
16. If q books cost £m, how many pence does 1 book cost?
17. Give a simpler answer for; $2p^2q \times 4q$.
18. Give a simpler answer for; $\frac{5a^2b}{10a}$.
19. Find p if $p - \frac{1}{2} = 3$.
20. Write down the cost of c pencils if 1 pencil costs x pence. Give your answer in £s.
21. What is the supplement of 23°?
22. What is the complement of $84\frac{1}{2}°$?
23. What name is given to an angle of 90°?
24. How many degrees are in a straight angle?

25. State whether the following angles are acute, obtuse or reflex: 23°, 31°, 154°, 302°, 5°.
26. If two angles of a triangle are 21° and 102° respectively, what size is the third angle?
27. If an isosceles triangle has one of its equal angles equal to 40°, what are the sizes of the other two angles of the triangle?
28. What size is each angle of an equilateral triangle?
29. Draw two straight lines and a transversal and show a pair of alternate angles.
30. Draw two straight lines and a transversal and show a pair of corresponding angles.
31. Add 20% of £1 to 10% of 10 pence. Answer in pence.
32. Add $\frac{1}{2}$ of $2\frac{1}{2}$ to $5 \times \frac{3}{5}$.
33. Express $(2\cdot01 - 1\cdot86)$ as a percentage.
34. Write in figures, two million and four.
35. Write 1001 in words.
36. Multiply 2·34 by 10 and add your answer to 7·8 ÷ 10.
37. What interest would be due on a loan of £10000 after one year at 15% simple interest?
38. Find the sum of 24, 326, 4028, 29 and 401.
39. Take 10% of £4 from 20% of £10. Answer in pence.
40. What fraction of 1 metre is 30 cm?
41. Write in simpler form: $12xy - xy + 3xy + x$.
42. Write in simpler form; $3x^2y \times 2x \times 4$.
43. Write in simpler form; $\frac{17a^2d}{34a}$.
44. Find d, if $3d = 81$.
45. Find t, if $\frac{t}{3} = 6$.
46. If a car travels x km in t hours, what rate, in km/h is it travelling?
47. Write down t% of £d.
48. Write in simpler form: $2t + 8t + 9t - 10t + 3$.
49. If $t = 3$, $p = 2$ and $q = 1$, find a number answer for $\frac{t^2 + p^2 + q}{21}$
50. Find x if $23x + 1 = 27 + 3$.
51. Find the value of x.

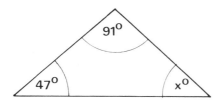

52.

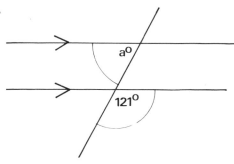

Find the value of *a*.

53.

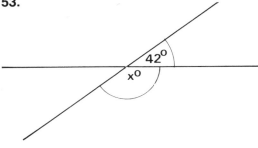

Find the value of *x*.

54.

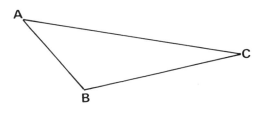

Name the longest side of △ABC.

55.

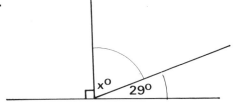

Find the value of *x*

56.

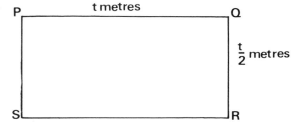

PQRS is a rectangle.
Find (a) the perimeter, in metres, of rectangle PQRS;

(b) the area, in square metres, of rectangle PQRS.

57.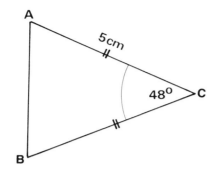

(a) State the length of BC.
(b) Find the size of ∠ABC.

58. What is the supplement of $p°$?

59. What is the complement of $t°$?

60.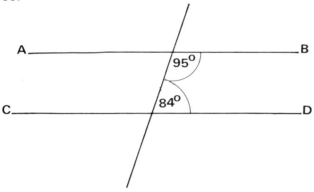

What can you say about the lines AB and CD?

61. Express as percentages: $\frac{2}{5}$; $\frac{1}{8}$ and $\frac{4}{5}$.

62. Find the average of 2, 3, 5, 9, 11 and 6.

63. What would have to be added to the sum of the six numbers in question **62** to give a total of 41?

64. A chair costing £129 is sold to gain 10% of what it cost. Find the selling price.

65. What should be taken from 11·29 to leave 7·345?

66. What must be added to $3\frac{1}{4}$ to make $7\frac{1}{8}$?

67. What percentage of £1 is 27 pence?

68. What fraction of £1 is 50 pence?

69. How many tiles, each a square of side length $\frac{1}{2}$ metre, will completely tile a floor which is rectangular, with a length of 8 metres and a width of $6\frac{1}{2}$ metres?

70. Arrange $\frac{1}{2}$, $\frac{3}{5}$, $\frac{7}{10}$, $\frac{11}{20}$ and $\frac{3}{4}$ in ascending order.

71. If $t = 5$ and $x = 4$, find a number answer for $2(t^2 + x)$.

72. Find p, if $3p - 2 = 7$.

76   A Basic Mathematics Workbook

73. Find a simpler way of writing $\dfrac{12a + 24a - 3a}{11}$

74. Write in simpler form $10a^2p \times 2ap^2$.

75. Write in simpler form $\dfrac{30x^2y}{10x}$.

76. If $a = 2$ and $c = 4$ write an equation connecting a and c.

77. What must be added to $2x$ to make $12x$?

78. What must be taken from $4xy$ to leave xy?

79. If $a = 7$ what is the value of $2a^2$?

80. Write in simpler form $12p - p - p - p$.

81. What is the supplement of $2p°$?

82. What is the complement of $5x°$?

83. 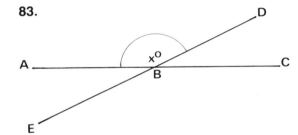 Name another angle in the figure which equals $x°$.

84. In the figure of question 83, what size is $\angle DBC$?

85.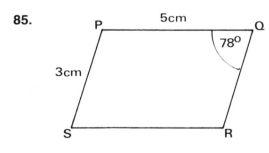

If PQRS is a parallelogram:
(a) what length is QR?
(b) what length is SR?
(c) what size is $\angle SPQ$?
(d) what size is $\angle PSR$?
(e) what size is $\angle SRQ$?

86.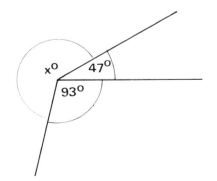

What size is x?

87.

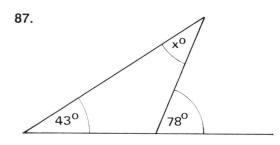

Find the value of x.

88.

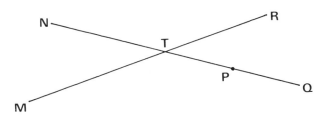

Name as many line segments as you can see in the figure opposite.

89.

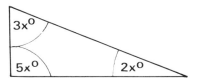

What would be the side length of a square of the same area as rectangle ABCD?

90.

Work out the value of x and then state the size of each angle of the triangle.

91. What must be added to half of $\frac{4}{9}$ to make 1?
92. Find 2% of half a million pounds.
93. VAT at 15% is added to a hotel bill of £250. What amount of VAT is added?
94. What is the least number that must be added to 133 to make it exactly divisible by 11?
95. Add 2 × 3·08 to $\frac{1}{3}$ of 9·6. Express your answer as a percentage.
96. Write, in figures, one quarter of a million.
97. What is the biggest number that divides evenly into 33 and 132?
98. Express 24·2 kg in mg.
99. Write, as decimal fractions $\frac{1}{2}, \frac{1}{4}, \frac{1}{8}$ and $\frac{1}{10}$.

78 A Basic Mathematics Workbook

100. Tom was born in 1972. How old will he be on his birthday in 2001?

101. Find x if $\dfrac{3x}{10} = 30$.

102. Write in simpler form $\dfrac{12mc^2}{6c}$.

103. Write in simpler form $4p \times 2q \times 8p$.

104. Express t metres in cm.

105. Express x cm in km.

106. Write in simpler form $13x - x + 11x$

107. If $a = 1$ and $b = 2$ give a number to represent $2a + 3b$.

108. Find t if $2t - 27 = 3$.

109. Find p if $p - 14 = 14$.

110. Write in simpler form $\dfrac{14p^2 d}{28d^2}$.

111. What is the complement of $2\tfrac{1}{2}°$?

112. What kind of angle is an angle of $102°$?

113.

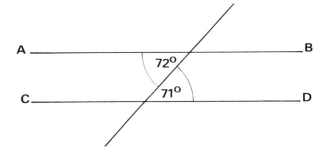

What can you say about the lines AB and CD?

114.

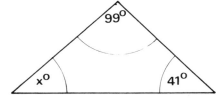

Find the value of x.

115.

Name a line which equals line BC + line CD.

116. What is the supplement of 142°?

117.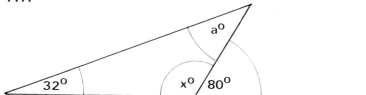

Find *a* and *x*.

118.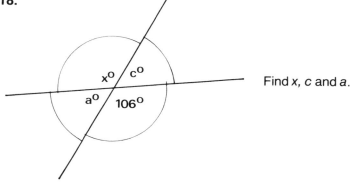

Find *x*, *c* and *a*.

119. Draw an isosceles triangle, ABC with AB=AC=4·5 cm and ∠BAC=38°. Measure BC.

120.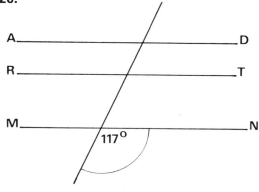

AD, RT and MN are parallel to one another.
Copy the figure and write in the sizes of all the un-marked angles.

121. Put into **descending** order $\frac{1}{2}$, $\frac{1}{3}$, $\frac{3}{4}$, $\frac{7}{8}$ and $\frac{5}{16}$

122. A 20% deposit is needed if an article costing £250 is to be bought on Hire Purchase. What deposit is needed?

123. A boy's age is one third that of his mother's. Their combined ages total 48 years. Find the age of each.

124. A jug of capacity $\frac{1}{2}$ litre is $\frac{2}{3}$ full. How many cm³ could it still hold?

125. Four cubes each have a side length of 2 cm. How many more identical cubes would give a total volume of 56 cm³?
126. Multiply 2 by $2\frac{1}{3}$ and subtract the result from 5.
127. Four boys are given £17·60 to divide equally among them. How much should each boy get?
128. What is the average of 1, 2, 3, 4, 5, 6 and 7?
129. Add 1·72 × 0·2 to 3 × 1·01.
130. Take the sum of $\frac{2}{3}$ and $\frac{4}{3}$ from 8.
131. Find p if $p + 18 = 35$.
132. If $a = 2$; What does $3a^2$ equal?
133. Write in simpler terms $5t - 2t + 3t - t + 8$.
134. By how much is $12a$ greater than $3a$?
135. If $t = 5$ and $q = 4$ find a number answer for $2t + q^2$.
136. Multiply $2pq$ by $3q$.
137. Divide $28a^2$ by $7a$.
138. If $c = 5$ what does $24c$ equal?
139. Write in simpler form $2ab + 12ab + 5ab + b$.
140. Find t if $\frac{t}{8} = 5$.
141. How many degrees equal the sum of two right angles and a straight angle?
142. What is the supplement of $t°$?
143. What is the complement of 32°?
144. Do parallel lines ever meet?
145.

If PQ and RT are parallel lines what size is ∠QPR?

146. What name is given to the pair of angles PRT and RPQ in the diagram of question **145**?
147. A right-angled triangle also has an angle of 45°. What kind of triangle must it be?

148.

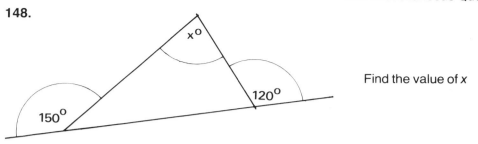

Find the value of x

149. One angle of a triangle is 20°. A second angle is $t°$. What size is the third angle?

150. One of the diagonals of a rectangle measures 10 cm. What is the length of the other diagonal?

151. What can you say about the three sides of a scalene triangle?

152.

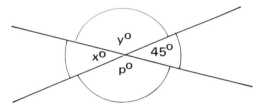

Find the value of x, y and p.

153.

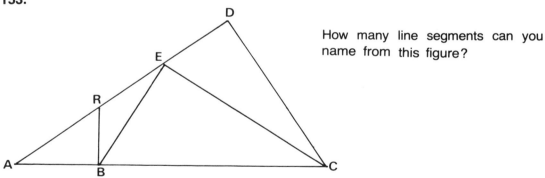

How many line segments can you name from this figure?

154. Write, in figures, the number twenty one thousand and six.

155. Write 2 000 004 in words.

156. A box of matches should contain 80 matches. If 10% of the matches have been used how many matches are left in the box?

157. What decimal fraction must be added to 0·079 to make 1?

158. Multiply 1·02 by 1000 and add 2·1 to your result.

159. What is the total surface area of the faces of a cuboid of length 3 metres, of breadth 2 metres and of height 2 metres?

82 A Basic Mathematics Workbook

160. What is the volume, in cm³ of the cuboid in the question above?

161. What percentage of 40 is 8?

162. What fraction of 14 is 7?

163. What percentage of 14 is 7?

164. $a+b=10$ if $a=b=5$: is this statement true or false?

165. If $t=4$ what is the value of $\frac{5}{2}t$?

166. a and b are two numbers such that a is greater than b by 7. If $b=6$, find a.

167. Find a if $2a-17=9$.

168. Give a number answer for $\frac{2a}{c}+\frac{a}{d}$ if $a=4, c=8$ and $d=1$.

169. Express t metres³ in cm³.

170. Express $2cd$ pence in £s.

171. Write in simpler form $3p+2p-p-3p+5p$.

172. The sum of $2a$ and $5a$ has $3a$ subtracted from it. What is the result?

173. If $a=4$ and $t=5$ what is the value of $2at^2$?

174. What is the supplement of $2p°$?

175.

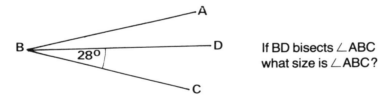

If BD bisects ∠ABC what size is ∠ABC?

176.

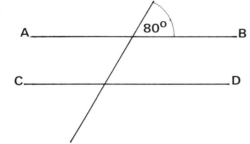

If AB and CD are parallel, how many angles of 80° are there in the figure?

177. One angle of a triangle is 41°. A second angle is 32°. What size must the third angle be?

178. What name is given to an angle which is smaller than 90°?

179. How many degrees are in a complete revolution?

180. What name is given to an angle greater than 180° but smaller than 360°?

181. Draw △ABC with AB = 4·8 cm, BC = 6·2 cm and CA = 5·3 cm. Measure ∠ABC.
182. Draw △PQR with PR = 4·2 cm, QR = 5·7 cm and ∠PRQ = 39°. Measure PQ.
183. What is the complement of $m°$?
184. Add $2\frac{1}{2} \times \frac{1}{3}$ to $\frac{1}{6}$ and take the result from 2.
185. A jug holds $\frac{3}{4}$ litre. If it is half full of water how many cm³ of water can it still hold?
186. Divide 2·04 by 0·02.
187. Multiply 4·2 by 0·01.
188. Add 1·2 × 10 to 0·03 × 100.
189. Take the sum of $1\frac{1}{2}$ and $3\frac{1}{2}$ from 10.
190. What percentage of 200 is 50?
191. What fraction of $4\frac{1}{2}$ is $1\frac{1}{2}$?
192. How many days are there in a leap year?
193. 1984 is a leap year and is divisible by 8. Are all leap years divisible by 8?
194. $c = 4$ and $t = 1$ What is the value of $2c + 3t$?
195. If $a = 5$, what is the value of a^3?
196. What value of x would make x^2 equal to x^3?
197. Find x if $2x - 3 = 109$.
198. Find x if $13x = 26$.
199. Find a if $\frac{a}{2} + 1 = 5$.
200. Find y if $\frac{y}{3} + 1 = 6$.
201. Find t if $2t = 4\frac{1}{2}$.
202. Write, in simpler form, $2m \times 3m^2 \times 6$.
203. Write, in simpler form, $\frac{5pm^2}{10m}$.
204. What angle is three times the size of a right angle?
205. How many degrees are in a straight angle?
206.

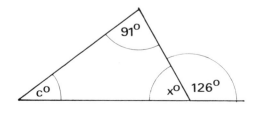

Find c and x.

207.

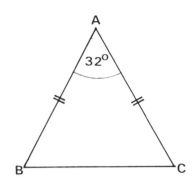

Find ∠ABC and ∠ACB.

208.

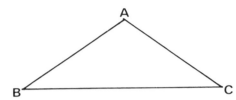

Why is the line BC smaller than the line BA added to line AC?

209. In the question above, if ∠b = 28° and ∠A is 111° what size is ∠C?

210.

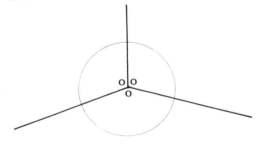

What size is each angle in the diagram?

211. What name is given to an angle of 121°?

212. If one side of an equilateral triangle measures 6 cm what must be the sum of the lengths of the other two sides?

213. One of the angles of an isosceles triangle measures 110°. What must be the sizes of the two other angles?

214. It costs 80p to go to a bingo session. 50 Ladies wish to go in a party to play bingo. For every 10 ladies, 1 is admitted free of charge. What is the total charge for admission?

215. A boy has £2.45. He wishes to buy 3 pencils costing 8p each and a ruler costing 24p. If he does this how much change will he have?

216. What fraction of 4 is 1?

217. What percentage of 4 is $\frac{1}{2}$?

218. What fraction of 8 is $\frac{1}{4}$?

219. Express 0·08 as a percentage.
220. What must be added to the sum of $\frac{1}{3}$ and $\frac{1}{6}$ to give 1?
221. What must be taken from 2·824 to leave 0·06?
222. What percentage of 102 is 306?
223. A car travels 87 km in 3 hours. At what rate is it travelling?
224. Find p if $\frac{2p}{3} = 4$.
225. Find a if $2a - 17 = 33$.
226. Multiply $14ab$ by $2a^2$.
227. Divide $4m^2n$ by n^2.
228. Add $2p$ and $7p$. From their sum, take the sum of $2p$ and p.
229. Write, in simpler form $2d + 8d - 6d + 4d - d + 5$.
230. A boy has £x. He gives away half his money and changes the rest to pence. How many pence has he?
231. Express $2c$ kg in mg.
232. At what speed is a car travelling if it covers $55d$ km in d hours?
233. Take the sum of $4x$ and $7x$ from the sum of $19x$ and $3x$.

234.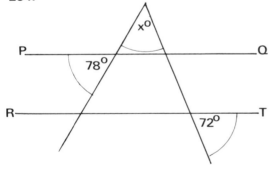
PQ is parallel to RT.
Find the value of x.

235.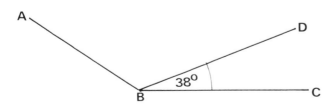
If $\angle$DBC is one third of $\angle$ABC, what size is $\angle$ABD?

236. What is the complement of $12t°$?
237. If $x°$ is the supplement of $y°$ write down an equation connecting x and y.
238. Draw a triangle whose angles are 32°, 49° and 99°. Is this triangle **unique**?

86 A Basic Mathematics Workbook

239. Draw △TMN with TM = 8 cm, MN = 6 cm and ∠TMN = 42°. Measure TN.

240. Draw a line 11·6 cm long. Call this line AB. Use your ruler and compasses to find the mid-point of AB **without actual measurement**.

241. If $2t°$ is the supplement of $p°$ write down an equation connecting t and p.

242. What name is given to an angle of 11°?

243.

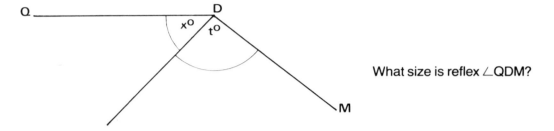

What size is reflex ∠QDM?

244. Find the sum of 31·2, 3·04 and 26·12.

245. Express as fractions in lowest terms: 0·8, 0·6, 0·08 and 0·002.

246. Express 0·41 as the sum of two fractions.

247. Express as decimals: $\frac{7}{10}$; $\frac{6}{100}$; $\frac{3}{1000}$; $\frac{64}{100}$; $\frac{235}{1000}$ and $\frac{7}{100}$

248. Divide $8\frac{2}{5}$ by $2\frac{5}{8}$.

249. Multiply $2\frac{2}{3}$ by $3\frac{1}{2}$ and divide your result by $2\frac{4}{5}$.

250. Express $3\frac{3}{8}$ as an improper fraction.

251. By a short method multiply 236 by 99.

252. The perimeter of a rectangle is 42·4 cm. One side is 5·1 cm long. Find the lengths of the other three sides.

253. A carpet measuring 4 m by 3 m is laid in a room with a floor area of 14 m². What area is left uncovered?

254. Find x, if $3x = 4\frac{1}{2}$.

255. Add $2a$ and $7a$ and take $3a$ from their sum.

256. If $t = 4$ what is the value of $5t^2$?

257. Multiply $4c$ by $2c^2$.

258. Divide $18mt^2$ by $9mt$.

259. Find c if $\frac{c}{2} + 1 = 3$.

260. Express t minutes in hours.

261. What is the number which is half of the number $14x$?

262. If $a = 2$ and $d = 3$ find a value for $2a + 7d$

263. If $a = 5$ and $t = 4$ and $xat = 40$ find x.

264. One angle of a triangle is 38°. A second is 102°. Find the third.

265.

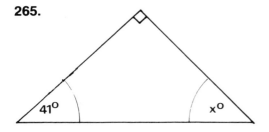

Find the value of x.

266.

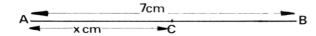

Write down the length of CB.

267.

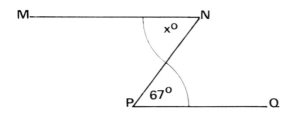

MN is parallel to PQ.
Find the value of x.

268.

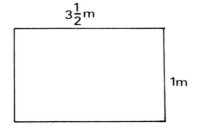

The figure shown is a rectangle.
Give its area in cm².

269. What is the complement of $\frac{1}{2} t°$?

270. What fraction of a straight angle is 20°?

271. What fraction of a right angle is $22\frac{1}{2}°$?

272.

If AB measures x cm and DB measures y cm write down the length of AD.

273. What name is given to an angle of 271°?
274. If 3 books each costing £1·75 are paid for from a £10 note what change will there be?
275. Multiply $2\frac{1}{4}$ by 8 and divide the result by $2\frac{1}{2}$.
276. Add 1·02, 2·03, 13·6 and 121.
277. Express $\frac{1}{4}$ of $\frac{1}{10}$ as a percentage.
278. Find the average of 8, 11, 17, 23 and 31.
279. What must be added to 1·2 × 2·4 to give 10·2?
280. Divide 2·04 by 0·02.
281. Multiply 23·1 by 0·01.
282. What interest is due on £100 borrowed for 6 months at 10% per annum simple interest?
283. Add 40% of £1000 to 20% of £3000.
284. Write, in simpler form; $24x - x + 2x + 9x + 7$.
285. Multiply $17pq$ by ca.
286. Divide $40x$ by 20.
287. Find p if $3p + 1 = 40$.
288. Express m days in hours.
289. Express t mg in kg.
290. Add $2x$, $4x$ and $11x$ and take $12x$ from your result.
291. Find m if $\frac{m}{4} = 2\frac{1}{2}$.
292. Write, in simpler form $\frac{12abc}{2bd} \times \frac{d}{ac}$.
293. What distance is travelled in p hours at x km/hour?
294. Draw $\triangle ABC$, with AB = 8·6 cm, BC = 9·8 cm and $\angle ABC = 29°$. Measure AC.
295. What is the complement of $3x°$?
296. What fraction of a complete revolution is 40°?

297.

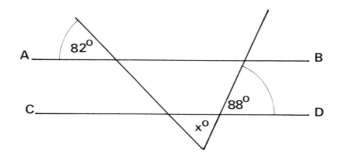

AB and CD are parallel lines.
What is the value of x?

500 Miscellaneous Questions

298. Draw an angle of 324°.
299. Draw a rectangle whose diagonals are 10 cm long. Is this rectangle **unique**?
300. If the angles of a triangle are $a°$, $x°$ and $t°$, write an equation connecting a, x and t.
301. A man bought a car for £4000 and sold it later at a loss of 25% on the cost price. How much money did he lose?
302. If 342 people out of 1000 do not like fish, what percentage is this?
303. How much greater than $2\frac{1}{4}$ is $3\frac{1}{2} \times 7$?
304. What interest is due to be paid on £2500 borrowed for 6 months at 12% per annum?
305. Which is smaller, $\frac{2}{5}$ or $\frac{4}{9}$?
306. What is the average of 14, 20, 22, 25 and 14?
307. An article costing £40 had its cost price increased by 40%. What was the new cost price?
308. How many centimetres are in 2·45 km?
309. An article costing £200 cash is sold on hire purchase for 40% deposit and 12 equal payments of £14·80. How much extra was charged for hire purchase?
310. How many litres of water can a cuboid hold if its internal measurements are 1·5 m, 1 m and 80 cm?
311. If $x = 2$ and $y = 3$ what is the value of $4(2x + y)$?
312. Find t if $2t + 1 = 11$.
313. How many cakes can be bought for £3 if p cakes cost x pence altogether?
314. Give a simpler answer for $2p + 8p - 7p + 6p$.
315. Find m if $3(m + 1) = 10$.
316. If m books cost £t how much does 1 book cost? Answer in pence.
317. Give a simpler answer for $2a^2 c \times 2c^2 a$.
318. Give a simpler answer for $\dfrac{18p^2 q}{20q}$.
319. Find t, if $t + 3 = 4\frac{1}{2}$.
320. Write down the cost of d pencils if t pencils cost m pence.
321. What is the supplement of 84°?
322. What is the complement of $2\frac{1}{2}°$?
323. What name is given to an angle of 180°?
324. How many degrees are in 4 right angles?
325. State whether the following angles are acute, obtuse or reflex: 49°, 181°, 274°.
326. If two angles of a triangle are 41° and 56° what size is the third angle?
327. If an isosceles triangle has one of its equal angles equal to 50° what are the sizes of the other two angles?

328. What size is each angle of a scalene triangle?
329. Draw 2 straight lines and a transversal and show a pair of corresponding angles.
330. Draw 2 straight lines and a transversal and show a pair of alternate angles.
331. Add 30% of £1 to 50% of £10.
332. Add $\frac{1}{4}$ of 10 to $2\frac{1}{2} \times 2$.
333. Express $(5 \cdot 68 - 2 \cdot 46)$ as a percentage.
334. Write in figures, two million three hundred thousand.
335. Write 1234 in words.
336. Multiply 2·1 by 100 and add your result to 42·6.
337. What interest would be due to a loan of £240000 for 1 month at a rate of 20% per annum?
338. Find the sum of 12, 121, 6 and 430.
339. Take 20% of £1 from 60% of £5.
340. What fraction of 1 metre is 20 cm?
341. Write in simpler form, $2xy + 12xy - 10xy$.
342. Write in simpler form, $2ab \times 14a^2b$.
343. Write in simpler form $\dfrac{40p^2q}{10pq^2}$.
344. Find x if $3\frac{1}{2}x = 7$.
345. Find t if $\dfrac{t}{4} = 14$.
346. If a car travels t metres in x seconds what rate is this in km/h?
347. Write down $p\%$ of £y.
348. Write in simpler form, $4x + 11x + 17x - 21x$.
349. If $t = 2$, $p = 1$ and $c = 3$ find the value of $t^2 + p^2 + c^2$.
350. Find x if $2x + 9 = 13$.
351.

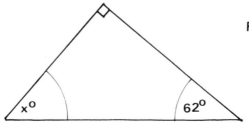

Find the value of x.

352.

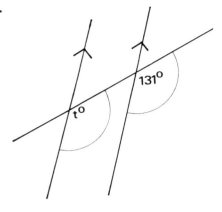

Find the value of t.

353.

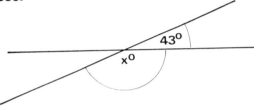

Find the value of x.

354.

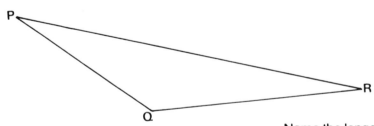

Name the longest side of △PQR.

355.

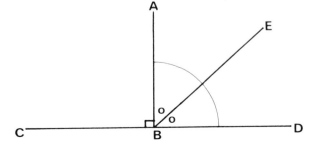

What size is ∠EBD?

356.

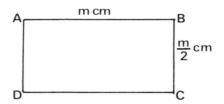

ABCD is a rectangle.
Find (a) the perimeter, in metres, of rectangle ABCD and (b) the area, in square cm, of rectangle ABCD.

357.

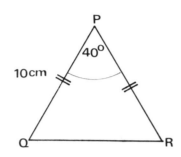

(a) What length is PR?
(b) What size is ∠PRQ?

358. What is the supplement of $2t°$?

359. What is the complement of $m°$?

360.

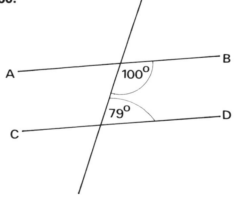

What can you say about the lines AB and CD?

361. Express as percentages: $\frac{1}{2}, \frac{1}{4}, \frac{3}{4}, 1$.

362. Find the average of 60, 70 and 20.

363. What would have to be added to the sum of the numbers in question 62 to give a total of 1000?

364. A table costing £60 was sold to gain 10% on what it had cost. What was it sold for?

365. What should be taken from 17·4 to leave 2·69?

366. What must be added to $11\frac{1}{4}$ to make 14?

367. What percentage of £4 is 50p?
368. What fraction of £10 is 50p?
369. How many tiles each $\frac{1}{2}$ metre long and $\frac{1}{2}$ metre wide will be needed to tile a rectangle, 10 metres long and 8 metres wide?
370. Arrange $\frac{5}{8}, \frac{7}{16}, \frac{13}{24}$ in ascending order.
371. If $a=4$ and $c=2$ find a value for $2a+3c$.
372. Find t if $4t=12$.
373. Find a simpler way of writing $\frac{3p+6p}{3}$.
374. Write in simpler form, $2xy+4x^2y$.
375. Write in simpler form, $\frac{8a^2b}{4ab}$.
376. If $a=4$ and $c=8$ connect a and c by writing an equation.
377. What must be added to $3c$ to make $11c$?
378. What must be taken from $20t$ to leave $2t$?
379. If $a=7$ what is the value of $\frac{2a^2}{14}$?
380. Write in simpler form $3p-p-p-p$.
381. What is the supplement of $3q°$?
382. What is the complement of $5t°$?
383.

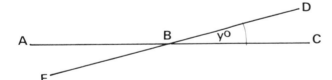

Name another angle in the figure which equals $y°$.

384. In the figure of question 83, what size is $\angle ABD$?
385. If $x=4$ and $y=2$ what is the value of $(x+y)^2$?
386.

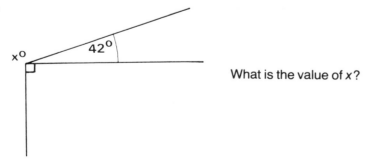

What is the value of x?

387.

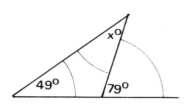

What is the value of x?

388.

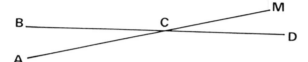

Name as many line segments as you can.

389.

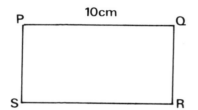

If the area of the rectangle is 40 cm² what length is QR?

390.

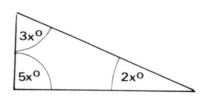

What size, in degrees, is each angle?

391. What must be added to half of $1\frac{1}{2}$ to make 10?

392. Find 10% of half a million.

393. V.A.T. at 15% is added to a bill of £300. How much is added?

394. What number is one quarter of one half of 4000?

395. Add $2 \times 1\cdot04$ to $3 \times 1\cdot12$.

396. Write half a million in figures.

397. What is the biggest number that divides evenly into 66 and 132?

398. Express $1\frac{1}{4}$ kg in g.

399. Write, as decimal fractions, $\frac{1}{8}$, $\frac{1}{4}$, $2\frac{1}{2}$.

400. Tom was born in 1973. How old will he be on his birthday in 2002?

401. Find p if $2\frac{1}{2}p = 7\frac{1}{2}$

402. Write in simpler form, $\dfrac{ac^2}{ca^2}$.

403. Write in simpler form, $2q \times 4q^2 \times 3q^3$.

404. Express t km in metres.

405. Express p cm in km.

406. Write in simpler form $2x + 7x - 3x - 6x$.

407. If $x = 4$ and $y = 10$, give a number to represent $2a + 5y$.

408. Find t if $3t - 1 = 14$.

409. Find p if $p - 17 = 1$.

410. Write in simpler form $\dfrac{2a}{4}$.

411. What is the complement of $2\frac{1}{2}x°$?

412. What kind of angle is an angle of 124°?

413.

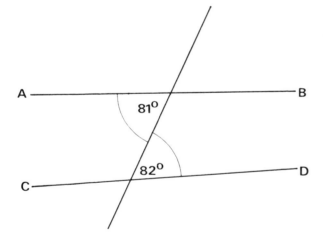

What can you say about the lines AB and CD?

414.

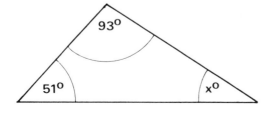

Find the value of x.

415.

Name a line which equals line BC + line CD.

416. What is the supplement of $3t°$?

417.

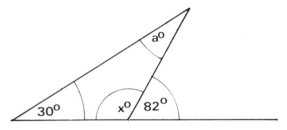

Find a and x.

418.

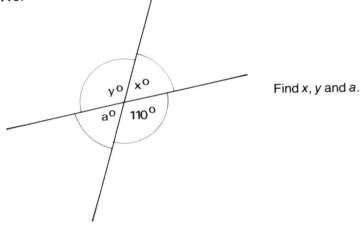

Find x, y and a.

419. Draw an isosceles triangle PQR with PQ = PR = 8 cm and $\angle$QPR = 60°. What length is QR?

420. Write in simpler form $3(2x + y) - 6x - 3y$.

421. Put into **descending** order: $\frac{1}{3}$, $\frac{3}{4}$, $\frac{1}{2}$, $\frac{7}{8}$ and $\frac{5}{16}$

422. A 10% deposit is needed on hire purchase to buy a TV set costing £542. How much of a deposit is needed?

423. A girl's age is one sixth of her mother's. The combined ages total 28. What age is each?

424. A jug of capacity $\frac{1}{4}$ litre is $\frac{1}{4}$ full. How many cm³ does it contain?

425. Six cubes each have a side length of 1 cm. What is their total volume?

426. Multiply $\frac{2}{3}$ by $1\frac{1}{8}$ and add your result to $2\frac{1}{2}$.

427. Four boys receive a total of £250 to divide among them equally. How much should each get?

428. What is the average of 2·4, 5·6, 7·8 and 1?

429. Add 1·2 × 2 to 4·6 × 4.

430. Take the sum of $\frac{2}{3}$ and $\frac{1}{4}$ from $2\frac{1}{4}$.

431. Find y if $3y - 4 = 8$.

432. If $x = 4$ what is the value of $\frac{1}{2}x^2$?

433. Write in simpler terms, $\frac{1}{2}x + \frac{1}{4}x + \frac{1}{3}x$.

434. By how much is $19xy$ greater than $2y$?

435. Multiply $\frac{1}{2}p^2$ by $\frac{1}{4}p^3$.

436. Multiply $4xy$ by $3x^2$.

437. Divide $104t^2$ by $8t$.

438. If $c = 20$ what is the value of $\frac{1}{4}c^2$?

439. Write in simpler form, $4ab + 6ab - ab$.

440. Find x if $\frac{x}{8} = 40$.

441. How many degrees equal the sum of 3 right angles and 2 straight angles?

442. What is the complement of $\frac{2}{3}y°$?

443. What is the supplement of 119°?

444. When do parallel lines meet?

445.

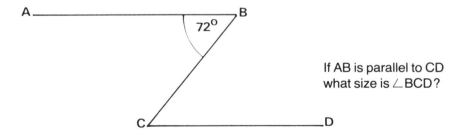

If AB is parallel to CD what size is ∠BCD?

446.

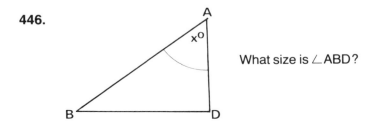

What size is ∠ABD?

447. A right-angled triangle also has an angle of 45°. What kind of triangle must it be?

448. What is the value of y?

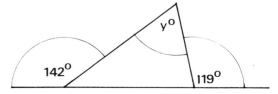

449. One angle of a triangle is 42°. A second is x°. What size is the third?

450. One of the diagonals of a rectangle equals 18 cm. What length is the other diagonal?

451. What can you say about the three sides of an equilateral triangle?

452. Find x, t, p.

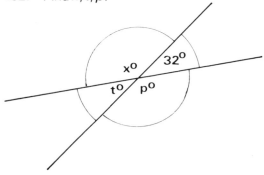

453.

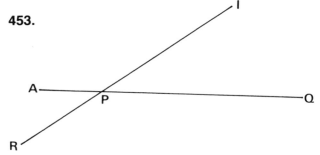

Name as many line segments as you can see in the figure.

454. Write, in figures, the number twenty two thousand and twenty two.

455. Write 2006 in words.

456. A box of matches should contain 90 matches. 10% is missing. How many are in the box?

457. What decimal fraction must be added to 1·46 to make 2?

458. Multiply 1·03 by 1000 and add 4·1.

459. What is the total surface area of a cuboid which measures 3 cm by 2 cm by 1 cm?

460. What is the volume, in cm^3, of the above cuboid?

461. What percentage of 200 is 50?
462. What fraction of 28 is 2?
463. What percentage of 30 is 10?
464. $a+b=20$ if $a=b=10$. True or false?
465. If $t=4\frac{1}{2}$ what does $\frac{5t}{2}$ equal?
466. a and b are two numbers such that a is one greater than b, If a is 74 find b.
467. Find c if $2c-17=15$.
468. If $x=4$ and $y=2$ find a value for $\frac{x}{y}+\frac{y}{x}$
469. Express x metres³ in cm³.
470. Express $4xy$ pence in £s.
471. Write in simpler form, $20xt+3xt-11xt$.
472. The sum of $2ab$ and $5ab$ is subtracted from 24. What is the result?
473. If $x=4$ and $t=5$ what is the value of $(x+t)^2$?
474. What is the supplement of $3p°$?

475.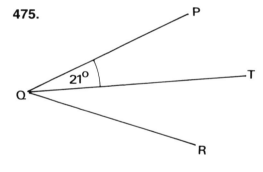

If QT bisects ∠PQR what size is ∠PQR?

476.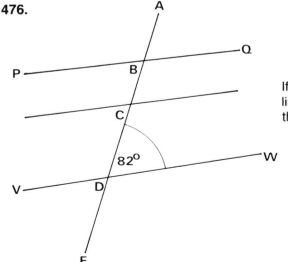

If PW, RT and VW are parallel lines how many angles of 98° are there in the figure?

477. One angle of an isosceles triangle is 104°. What size is each of the other two?
478. What name is given to an angle whose size lies between 90° and 180°?
479. How many degrees are in 4 complete revolutions?
480. What name is given to an angle which lies between 180° and 360°?
481. Draw $\triangle$ PQR with PQ = PR = 6·6 cm and $\angle$P = 60°. What length is QR?
482. Draw $\triangle$ PQR with PQ = QR = 4·8 cm and $\angle$Q = 90°. What size is $\angle$P?
483. What is the complement of $mn°$?
484. Add $\frac{1}{2}$ of $\frac{1}{4}$ to $\frac{1}{8}$ and take your answer from 1.
485. A jug holds 2 litres. If it is a quarter full how many cm³ could it still hold?
486. Divide 2·65 by 0·02.
487. Multiply 2·46 by 0·001.
488. Add 1·2 × 2 to 3·4 ÷ 2.
489. Take the sum of $1\frac{1}{2}$ and $2\frac{3}{4}$ from 5.
490. What percentage of 300 is 10?
491. What fraction of $2\frac{1}{4}$ is $1\frac{1}{8}$?
492. How many days are in a leap year?
493. Are all leap years divisible by 8?
494. $d = 1$ and $m = 6$. What is the value of $(2d + m)^3$?
495. If $c = 4$ what does c^3 equal?
496. What value of c would make $2c^3$ equal to 2?
497. Find x if $3x + 1 = 4$.
498. Find d if $2d^2 = \frac{1}{2}$.
499. Find n if $\frac{n}{2} = 4$.
500. Find y if $\frac{y}{4} + 1 = 6$.